31st International Symposium on Space Terahertz Technology (ISSTT 2020)

Tempe, Arizona, USA
8 - 11 March 2020

ISBN: 978-1-7138-2944-7

Printed from e-media with permission by:

Curran Associates, Inc.
57 Morehouse Lane
Red Hook, NY 12571

Some format issues inherent in the e-media version may also appear in this print version.

Printed with permission by Curran Associates, Inc. (2021)

International Symposium on Space Terahertz Technology
c/o NRAO Headquarters
520 Edgemont Road
Charlottesville, VA 22903-2475
USA

Phone: (434) 296-0254
Fax: (434) 296-0278

mbishop@nrao.edu

Additional copies of this publication are available from:

Curran Associates, Inc.
57 Morehouse Lane
Red Hook, NY 12571 USA
Phone: 845-758-0400
Fax: 845-758-2633
Email: curran@proceedings.com
Web: www.proceedings.com

Conference Schedule

SUNDAY, MARCH 9 2020

5pm - 7:30pm - **Reception**, Four Peaks Wilson tasting room

MONDAY, MARCH 9 2020

8:00 - **Coffee and Registration**

8:45 - **Welcome**, Nancy Gonzales - Dean of Natural Sciences, The College of Liberal Arts and Sciences, Arizona State University

Session I: Suborbital

Chair: Chris Groppi

10:30 - **Coffee Break**

Session II: Missions and Concepts

Chair: Jose Siles

12:00 - ***Submillimeter Wave Differntial Absorption Radar For Water Vapor Sounding In The Martial Atmosphere***, Omkar Pradhan, Ken Cooper, Leslie Tampari, Brian Drouin, Raquel Monje, Richard Roy, Jose Siles and Corey Cochrane.....8

12:20 - ***Millimetron Space Observatory Mission Development***, Andrey Baryshev, Andrey Smirnov, Evgeny Golubev, Mikhail Arkhipov, Elena Filina, Victor Pyshnov, Nelly Myshonkova, Sergey Fedorchuk, Igor Vinogradov, Thijs Graauw De and Sergey Likhachev.....9

12:40 - Lunch Break

2:00 - Invited Talk - *Submillimeter Wave Power Generation with InP HEMT Technology*, Bill Deal.....10

Session III: Schottky Diode and CMOS receivers

Chair: Imran Mehdi

2:30 - *Technology Roadmap for the HEterodyne Receiver for Origins (HERO)*, Martina Wiedner.....11

2:50 - *High-Sensitivity Terahertz Detection Module (HSTDM) onboard China's Space Station*, Sheng-Cai Shi.....12

3:10 - *A Single Sideband 530-600 GHz Integrated Receiver Utilizing Tunable Waveguide Filters*, Theodore Reck, Daniel Koller, Jeffrey Hesler and Eric Bryerton.....13

3:30 - Coffee Break

3:50 - ***MetOp-SG Ice Cloud Imager 183–664 GHz Front-End Engineering and Qualification Model Test Results***, Patrick Pütz, Bertrand Thomas, Michael Brandt, Guido Sonnabend, Tobias Stangier, Pia Krause, Ralf Henneberger, Monica Trasatti, Martin Philipp, Andreas Kilian, Hugh Gibson, Simon Rea, Hui Wang, Manu Henry, Chris Howe, Kai Parow-Souchon, Brian Moyna, Brian Ellison, Ana Andrés-Beivide, Marc Bergada, Noelia Alcaraz, Jaione Martinez, Michael Gotsmann and Ulf Klein.....N/A

4:10 -*180 GHz CMOS Pulsed Transmitter and Heterodyne Receiver Pair for in-situ Chemical Detections*, Deacon Nemchick, Brian Drouin, Adrian Tang, Yanghyo Kim, Maria Alonso and M.C. Frank Chang.....14

4:30-4:50 - *Integrated Silicon Platform for Co-planar Design of Vertically Stacked 2.06 THz Mixer Module*, Christine Chen.....15

5:45 - Buses Leave ISTB4 to Botanical Garden - Banquet

6:00 - Reception Banquet, Desert Botanical Garden

TUESDAY, MARCH 10 2020

8:00 - Coffee and Registration

8:45 - Invited Talk - *FIR Astrophysics in the US*, Kartik Sheth.....N/A

Session IV: Novel Devices

Chair: Boris Karasik

9:15 - *Heterodyne receiving with frequency combs: towards simultaneous ultrabroadband spectroscopy*, David Burghoff.....17

9:35 - *Wideband Superconducting Parametric Amplifiers for Millimeter-wave Instruments*, Peter Day, Nikita Klimovich, Byeongho Eom and Henry Leduc.....18

WEDNESDAY, MARCH 11 2020

8:00 - Coffee and Registration

8:45 - Invited Talk - *Steward Observatory's Radio Telescopes: Recent Advancements and Future Developments*, Alyson Ford.....N/A

Session VII: Optics and Waveguide Components

Chair: Andrey Baryshev

9:15 - ***4 THz beam filter based on a back to back Si-lens system***, Yuner Gan, Behnam Mirzaei, Sebastiaan van der Poel, Jose Silva, Matvey Finkel, Martin Eggens, Marcel Ridder, Ali Khalatpour, Qing Hu, Floris van der Tak and Jian Rong Gao.....42

9:35 - ***Silicon micromachined receiver calibration switch for THz frequencies***, Adrian Gomez Torrent, Umer Shah and Joachim Oberhammer.....43

9:55 - ***Characterization of Dielectric Material at 300 GHz for Vacuum Window Applications***, C. Edward Tong, Keara Carter and Jake Connors.....44

10:15 - ***Contactless rotating MEMS waveguide switch for water detection at 557 GHz***, Sofia Rahiminejad, Cecile Jung-Kubiak, Mina Rais-Zadeh and Goutam Chattopadhyay.....47

10:35 - Coffee Break

11:10 - ***Receivers for the wideband Submillimeter Array***, Paul Grimes, Raymond Blundell, Scott Paine, Edward Tong and Lingzhen Zeng.....48

Session VIII: SIS Mixers and Receivers

Chair: Edward Tong

11:30 - ***A Horn-coupled 4-beam Dual-polarization Balanced SIS Mixer Based on Planar-integrated Circuits***, Wenlei Shan, Shohei Ezaki, Keiko Kaneko, Akihira Miyachi, Takafumi Kojima and Yoshinori Uzawa.....55

11:50 - ***Advanced Tuning Algorithms for High-Frequency SIS Mixers***, Ronald Hesper, Jan Barkhof, Andrey Baryshev, Tobias Vos, Giorgio Siringo, Neil Phillips and Pavel Yagoubov.....56

12:10 - ***Optics and Feed Design for the wSMA Receiver System***, Paul Grimes, Scott Paine, Lingzhen Zeng and Edward Tong.....57

12:30 - Lunch Break

2:00 - Invited Talk - *The Evolution of Antenna Technology: Past, Present and Future,* Constantine Balanis.....N/A

2:30 - ***Terahertz MgB2 HEB mixers with a 13GHz gain bandwidth***, Narendra Acharya, Evgenii Novoselov and Sergey Cherednichenko.....65

2:50 - ***Experimental Characterization of the LO Heating Effect in THz SIS mixers***, Alessandro Traini, Boon-Kok Tan, Ghassan Yassin, John Garrett, Andrey Khudchenko, Ronald Hesper, Andrey Baryshev and Valery Koshelets.....66

3:40-4:40 - Meteorite Vault and lab Tours - ISTB4

Poster Presentations

Proceedings of the 31st Symposium on Space Terahertz Technology:

ISSTT 2020

8-11 March 2020

Tempe, Arizona

Hosted by:

ASU School of Earth and Space Exploration
Arizona State University

The 2020 31st IEEE International Symposium on Space Terahertz Technology (ISSTT 2020) took place in Tempe, Arizona, USA on March 8-11, 2020. It was held on the grounds of Arizona State University and organized by the local committee consisting of engineers and scientists from ASU.

Held just before COVID-19 began to severely restrict travel to and from the USA, attendance at this conference was impacted. But nonetheless, the conference successfully and safely hosted 88 registered participants from 8 countries, with 80 papers spread over 8 oral sessions and 1 poster session presented. The option to both present papers and attend the conference remotely by zoom was available to all who could not attend in person. Of all the papers, four invited talks were given by prominent scientists, with some of them leading new large missions. The Symposium presentations covered a broad range of topics including instruments, receivers, THz mixers and local oscillators, large detector array, optics, calibration techniques, and some new ideas related to device physics. A full symposium agenda and list of presentations can be found at the Symposium website: http://www.isstt2020.com.

Special thanks to:

Virginia Diodes, Inc.

ASU Ira A. Fulton Schools of Engineering
Arizona State University

ASU NewSpace
Arizona State University

ASU School of Earth and Space Exploration
Arizona State University

GROPPI LABS
TERAHERTZ DESIGN

Local Organizing Committee

Chris Groppi

Phil Mauskopf

Sean Bryan

Georgios Trichopoulos

Saeed Zeinolabedinzadeh

Scientific Advisory Committee

Victor Belitsky (Chalmers University of Technology, Sweden)

Andrey Baryshev (University of Groningen, The Netherlands)

Brian Ellison (Rutherford Appleton Laboratory, United Kingdom)

Jian-Rong Gao (SRON, The Netherlands)

Gregory Goltsman (Moscow State Pedagogic Institute, Russia)

Christopher Groppi (Arizona State University, USA)

Jeffrey Hesler (Virginia Diodes Inc., USA)

Heinz-Wilhelm Hübers (DLR, Germany)

Boris Karasik (Jet Propulsion Laboratory, USA)

Valery Koshelets (IRE RAS, Russia)

Alain Maestrini (Paris Observatory, France)

Hiroshi Matsuo (NAOJ, Japan)

Imran Mehdi (Jet Propulsion Laboratory, USA)

Patricio Mena (University of Chile, Chile)

Netty Honingh (University of Cologne, Germany)

Christophe Risacher (IRAM, France)

Sheng-Cai Shi (Purple Mountain Observatory, China)

Jan Stake (Chalmers University of Technology, Sweden)

Edward Tong (Smithsonian Astrophysics Observatory, USA)

Yoshinori Uzawa (NAOJ, Japan)

Ghassan Yassin (University of Oxford, United Kingdom)

The Best PhD Student ISSTT2020 Contribution Award Committee

Jeffrey Hessler, Chairman Virginia Diodes Inc., USA

Urs Graf, University of Cologne, Germany

Victor Belitsky, Chalmers University of Technology, Sweden

Winner of the best PhD Student ISSTT2020 Contribution

Development of a Dual-Band Metamaterial Lens for Cubesat Water Observation, Cassandra Whitton

The Gal/Xgal Ultra-Long Duration Balloon-borne Spectroscopic THz Observatory (*GUSTO*)

C. K. Walker[1], C. A. Kulesa[1], Paul Goldsmith[2], and the *GUSTO* Team

The Gal/Xgal Ultra-Long Duration Balloon-borne Spectroscopic THz Observatory (GUSTO) will dramatically improve our understanding of the Universe by probing the topology and ecology of interstellar gas throughout the Milky Way and nearby galaxies. GUSTO is a balloon-borne, 0.9 m Cassegrain telescope with cryogenic heterodyne array receivers designed to stay aloft for 100 days or more. During this time GUSTO will survey 124 square degrees of the Milky Way and all of the Large Magellanic Cloud (LMC) in three important interstellar lines: [CII], [OI], and [NII] at 158, 63, and 205 μm, respectively. GUSTO will map the structure, dynamics, energy balance, pressure, and evolution of the Interstellar Medium within the Milky Way and LMC. GUSTO is an Explorer Mission of Opportunity. The mission will utilize the 100+ day flight potential of the Super Pressure Balloon, also known as the Ultra Long Duration Balloon (ULDB), provided by NASA's Balloon Program Office. GUSTO features a proven measurement approach, a high-heritage payload, and a simple, repeatable observing strategy that, combined with the ultra-long duration capability of the SPB, enables these important new galactic/extragalactic observations at a fraction of the cost of a comparable orbital mission

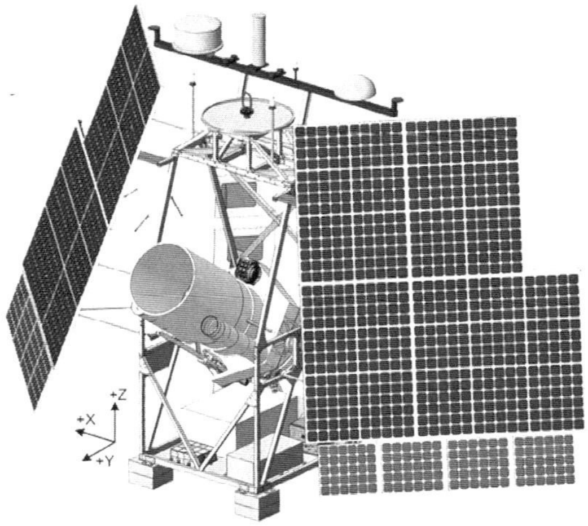

GUSTO ballooncraft and payload include a 0.9 m telescope, cryogenic THz receivers, and a gyro stabilized pointing system to provide an unparalleled galactic-extragalactic survey capability.

[1] Steward Observatory, 933 N. Cherry Ave., University of Arizona, Tucson AZ 85721 USA.

[2] Jet Propulsion Laboratory, 4800 Oak Grove Dr., Pasadena CA 91109

First Flight of an Almost All-CMOS 183 GHz Limb-Sounding Spectrometer System Aboard the ReckTangLEAllooncraft

Adrian Tang[12], Deacon Nemchick[1], Maria Alsono[1], Goutam Chattopadhyay[1], Theodore Reck[3], Yan Zhang[2], Yangyho Kim[4], Frank Chang[2]

This abstract discusses a compact and low-power 183 GHz heterodyne spectrometer instrument that is almost entirely implemented in CMOS technology except for a single InP preamplifier MMIC. The 1.1W and 0.5 Kg spectrometer system is intended for both Earth and planetary science applications enabling limb-sounding of H_2O from compact CubeSat and SmallSat platforms.

The spectrometer's RF front-end employs an 28nm system-on-chip (SoC) device containing a 183 GHz receiver, integrated 183 GHz phase-locked loop to provide the LO, integrated IF amplification and an integrated digital processor connected to many sensors and tuning knobs (bias and varactors) for calibration purposes. External to the receiver SoC a single 35nm InP MMIC low-noise amplifier (LNA) is placed in front of the CMOS receiver to maintain reasonable receiver noise temperatures. T_{sys} of the CMOS receiver alone is above 2000K while T_{sys} with the I_nP amplifier is on the order of 500K when operated at room temperature. A second 65nm CMOS chip provides the back-end processing for the spectrometer instrument. This chip contains a high-speed 3b analog-to-digital converter (ADC) and a 4096-channel wideband FFT processor, and SRAM based accumulator along with several sub-processors that perform analog calibration and timing adjustment tasks on the ADC. The chip also contains a USB2.0 interface allowing it to directly interface with a PC or spacecraft computer. While the ADC is capable of operation up to 6 GS/s it was operated at 1.5 GS/s in this instrument as the CMOS receiver's integrated IF only offers 750 MHz of bandwidth.

The full CMOS spectrometer instrument was integrated into the payload of a small hand-launched high-altitude ballooncraft called the "Reck-Tang Limb-sounding Experiment" (ReckTangLE) which was launched on Oct 17 2019 and performed several spectroscopic soundings of H_2O in the upper stratosphere over the southwestern United States.

[1]Jet Propulsion Laboratory, Pasadena California [2]Electrical and Computer Engineering, University of California Los Angeles [3] Virginia Diodes Inc. [4] Naval Research Laboratory.

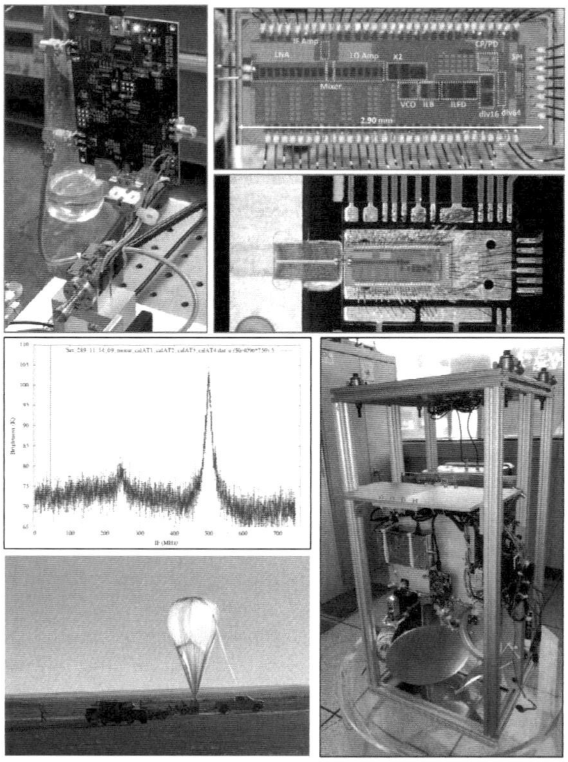

Fig. 1. Overview of the 183 GHz almost all-CMOS spectrometer instrument and integration on board a ballooncraft.

References

[1] Y. Kim, Y. Zhang, T. Reck, D. Nemchick, G. Chattopadhyay, B. Drouin, M-C F. Chang and A. Tang, "A 183-GHz InP/CMOS-Hybrid Heterodyne-Spectrometer for Spaceborne Atmospheric Remote Sensing," in IEEE Trans. on THz Sci. and Tech., vol. 9, no. 3, pp. 313-334, May 2019.

[2] A. Tang, Y. Kim, T. Reck, G. Chattopadhyay, I. Mehdi, B. Drouin, K. Cooper, N. Livesey M-C F. Chang,"DDFS and ΣΔ Approaches for Fractional Frequency Synthesis in Terahertz Instruments", IEEE Trans. on THz Sci. and Tech, Vol. 8 No. 4, pp 410-417, August 2018.

[3] Yan Zhang, Yanghyo Kim, Adrian Tang, Jonathan H. Kawamura, Theodore J. Reck, Mau-Chung Frank Chang "Integrated Wide-Band CMOS Spectrometer Systems for Spaceborne Telescopic Sensing", IEEE Trans. on Circuits and Systems I, Vol. 66, No.5, pp. 1863-1873, May 2019.

The Terahertz Intensity Mapper (TIM): Far-infrared Balloon Mission for Spectroscopic Galaxy Evolution Studies

R.M.J. Janssen[1,2], J.E. Aguirre[3], P.S. Barry[4,5], J. Bracks[3], C.M. Bradford[1,2], B. Bumble[1], A.J. Corso[3], J.P. Filippini[6], C.E. Groppi[7], D.P. Marrone[8], M. Bethermin[9], T.-C. Chang[1,2], M.J. Devlin[3], O.P. Dore[1,2], J. Fu[6], S. Hailey-Dunsheath[2], J. Hoh[7], G.P. Holder[6], G. Keating[10], H.G. LeDuc[1], R.P. Keenan[7], E.D. Kovetz[11], G. Lagache[9], L. Liu[2], H. Mani[7], J. Mathewson[7], R. Nie[6], P. Mauskopf[7], D. Narayanan[12], G. Popping[13], J. Redford[2], E. Shirokoff[14], A.K. Sinclair[7], R.S. Somerville[15], I. Trumper[16], M. Underhill[7], B. Uzgil[13], J.D. Vieira[6], and J. Zmuidzinas[2]

Understanding the formation and evolution of galaxies over cosmic time is one of the foremost goals of astrophysics and cosmology today. The cosmic star formation rate has undergone a dramatic evolution over the course of the last 14 billion years, and dust obscured star forming galaxies (DSFGs) are a crucial component of this evolution. A variety of important, bright, and unextincted diagnostic lines are present in the far-infrared (FIR) which can provide crucial insight into the physical conditions of galaxy evolution, including the instantaneous star formation rate, the effect of AGN feedback on star formation, the mass function of the stars, metallicities, and the spectrum of their ionizing radiation.

TIM, the Terahertz Intensity Mapper, is a NASA balloon mission that will observe the universe in the crucial gap between the spectroscopic coverage of the Atacama Large Millimeter/submillimeter Array (ALMA) in the sub/mm, and the James Webb Space Telescope (JWST) in the mid-IR; something which is impossible to from the ground. TIM will survey two 0.1 square degree fields centered on GOODS-S and the South Pole Telescope (SPT) Deep Field. TIM will produce deep maps of the 3D structure of the Universe by redshift tomography ("intensity mapping") with [CI], and [CII] X [NII] cross-spectra, to constrain the cosmic star formation history at cosmic noon. In addition, it is expected to achieve spectroscopic line detections of ~100 galaxies in the atomic fine structure lines of C, N, and O, as well as establish mean galaxy properties such as star formation rate, metallicity and AGN content, using a stacking analysis of known sources and the wealth of ancillary data available in the GOODS-S and SPT Deep Field.

To achieve these science goals, TIM will fly two longslit (1 degree slit length) grating spectrometers, which cover the 240-317 um and 317-420 um wavelength bands at R~250. Each of these spectrometers is serviced by a ~4000 pixel array of horn-coupled kinetic inductance detectors (KIDs). The KIDs, lumped-element resonator of superconducting Al, are designed to achieve a photon noise limited performance at 100 fW of loading with an absorption efficiency of >80%. They will be read out using an RFSoC based readout system.

The TIM gondola and cryogenics will be based upon proved BLAST hardware. However, a new 2-meter low-emissivity high-throughput carbon fiber telescope will be used to achieve maximum mapping speed.

The in-flight demonstration of both the high-sensitivity and scalable KID arrays as well as the low-emissivity high-throughput mirror are key technical milestones to future space-borne instrumentation such as the Origins Space Telescope (OST, formerly the Far-IR Surveyor) or Probe mission.

We will present the design of these key components for TIM as well as summarize the planned route for experimental development and testing, which will conclude with a launch from Antarctica in 2024.

[1] NASA Jet Propulsion Laboratory, Pasadena, CA 91109 USA
[2] Department of Astronomy, California Institute of Technology, Pasadena, CA 91125 USA
[3] Department of Physics and Astronomy, University of Pennsylvania, Philadelphia, PA 19104 USA
[4] High-Energy Physics Division, Argonne National Laboratory, Argonne, IL 60439 USA
[5] Kavli Institute for Cosmological Physics, University of Chicago, Chicago, IL 60637 USA
[6] Department of Astronomy, University of Illinois Urbana-Champaign, Urbana IL 61801 USA
[7] School of Earth and Space Exploration, Arizona State University, Tempe, AZ 85287 USA
[8] Steward Observatory, University of Arizona, Tucson, AZ 85721 USA
[9] Laboratoire d'astrophysique de Marseille, 13388 Marseille FRANCE
[10] Harvard Smithsonian Center for Astrophysics, Cambridge, MA 02138 USA
[11] Department of Physics and Astronomy, Johns Hopkins University, Baltimore, MD 21218 USA
[12] Department of Astronomy, University of Florida, Gainsville, FL 32611 USA
[13] Max Planck Institute for Astronomy, 69117 Heidelberg, Germany
[14] Department of Astronomy and Astrophysics, University of Chicago, Chicago, IL 60637 USA
[15] Department of Physics and Astronomy, Rutgers University, Piscataway, NJ 08854 USA
[16] Intuitive Optical Design Lab LLC, Tucson, AZ, USA

NOTES:

In-flight performance of the BLAST-TNG Kinetic Inductance Detector arrays and Readout Electronics

A. Sinclair[1], S. B. Gordon[2], P. Ade[3], P. Ashton[4], H.-M. Cho[6], G. Coppi[5], A. Corso[5], M. Devlin[5], S. Dicker[5], B. Dober[6], L. Fissel[7], N. Galitzky[8], J. Gao[6], C. Groppi[1], G. C. Hilton[6], J. Hubmayr[6], K. Irwin[9], J. Klein[5], D. Li[8], N. Lourie[9], I. Lowe[5], H. Mani[1], P. D. Mauskopf[1], C. McKenney[6], F. Nati[10], G. Novak[11], E. Pascale[12], G. Pisano[3], J. Soler[13], C. Tucker[3], M. Underhill[1], M. Vissers[6], C. Wheeler[14], P. Williams[11]

The BLAST-TNG experiment consists of a 2.5 meter Cassegrain telescope that feeds a cryogenic three-color submm imaging polarimeter [1],[2]. The receiver contains three arrays of superconducting kinetic inductance detectors (figure 1) observing in photometric bands centered at wavelengths of 250, 350 and 500 μm. The detectors are coupled to circular waveguides that feed profiled horn antennas. Each horn couples to two detectors designed to absorb radiation in orthogonal polarizations.

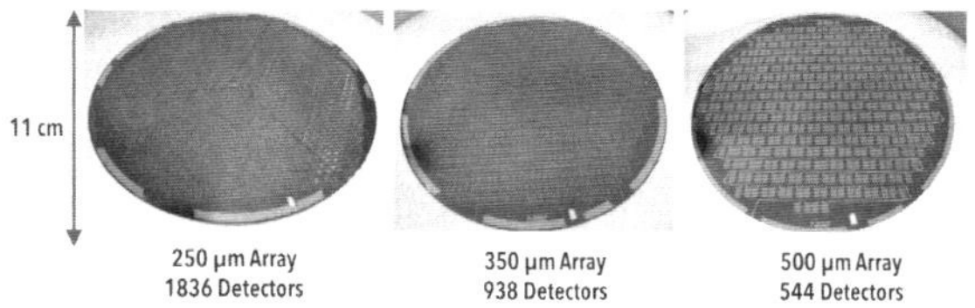

Fig. 1. The three arrays of kinetic inductance detectors in the BLAST-TNG receiver.

The readout electronics consisted of 5 ROACH2 FPGA boards with ADC/DAC boards from Techne Instruments. The electronics are custom modified for the balloon environment by the use of heat pipes to cool the FPGA chips and microprocessors [3].

BLAST-TNG had a stratospheric flight from Antartica in January, 2020. During the line of sight period of operation, we performed diagnostics of the receiver. We were able to read out approximately 3000 detectors over the three arrays.

Fig. 2. BLAST-TNG ready for launch.

The readout electronics worked well in flight and we monitored the temperature at float. We will report on the results of the receiver calibration from flight skydips and observations of bright submillimeter sources.

References

[1] N. Lourie, et al., "Preflight characterization of the BLAST-TNG receiver and detector arrays," *Proc. SPIE,* 10708, 2018..

[2] N. Galitzki, et al., "The next generation BLAST experiment," *Journal Astr. Instr..*, vol. 3, pp. 1440001, 2014.

[3] S. Gordon, et al., "An Open-Source FPGA-based LeKID readout for BLAST-TNG: Pre-flight results," *Journal Astro. Instr.,* vol. 5, 1641003, 2016.

[1] School of Earth and Space Exploration, Arizona State University, Tempe, AZ 85281

[2] STR, Woburn, MA, 01801

[3] School of Physics and Astronomy, Cardiff University, Cardiff, UK CF24 3AA

[4] Lawrence Berkeley Laboratories, Berkeley, CA.

[5] Department of Physics and Astronomy, University of Penn., Philadelpiha, PA, 19104

[6] NIST, Boulder, CO, 80305

[7] Department of Physics, Engineering Physics and Astronomy, Queen's University, Kingston, CA, K7L 3N6

[8] Stanford University, Palo Alto, CA, 94305

NOTES:

ASTHROS - Astrophysics Stratospheric Telescope for High-Spectral Resolution Observations at Submillimeter-waves: Mission Overview and Development Status

Jose V. Siles[1], Jorge Pineda[1], Jonathan H. Kawamura[1], Cristopher Groppi[2], Pietro Bernasconi[3], Joshua Gundersen[4], and Paul F. Goldsmith[1].

The Astrophysics Stratospheric Telescope for High-spectral Resolution Observations at Submillimeter-wavelengths, ASTHROS, is a 2.5-m (SOFIA-like size) balloon-borne observatory that will make the first detailed spectrally-resolved high spatial resolution 3D map of ionized gas in Galactic and extragalactic star forming regions via simultaneous observations of the 122μm (2.459 THz) & 205μm (1.461 THz) fine structure lines of ionized nitrogen.

ASTHROS builds on the success of heterodyne instruments on Herschel and SOFIA, providing a low-risk low-cost stepping stone for future heterodyne missions. A 21-day Antarctic flight in 2023 will focus on mapping two template Galactic star forming regions and the entire disk of the M83 barred spiral galaxy at high angular resolution, complementing existing datasets from SOFIA, WISE, Herschel, Spitzer and HST. ASTHROS will be capable of tuning to nearby spectral lines (OH, HDO, HF, HD, CO) for Target of Opportunity observations. One compelling target is the HD 112μm (2.674 THz) line that traces the gas mass in protoplanetary disks. ASTHROS' angular resolution of 12" at 122 μm and 20" at 205 μm corresponds to 0.2 pc and 0.35 pc at 122μm and 205μm, respectively, for a source 4 kpc from the Sun. This high angular resolution will enable us to resolve structures ~750 times smaller than the typical size of star forming regions (~150 pc). Combined with large-scale mapping, we will begin to understand how different stellar feedback mechanisms affect ionized gas over a wide range of spatial scales in the Milky Way and the M83 galaxy.

ASTHROS payload will consists of a 4-pixel dual band cryogenic superconducting heterodyne array camera for high-spectral resolution imaging at 1.4-1.5 THz and 2.4-2.7 THz. The instrument design features a straightforward receiver architecture, simple optical layout, and subsystems that have high degree of flight heritage, pedigree, and proven performance through suborbital and space missions such as STO-2 and Herschel. ASTHROS will fly for the first time a 4-K class low-power cryocooler and thus will not require liquid helium. A cryocooler will enable extended lifetime missions, and its use will serve as a pathfinder for future NASA missions.

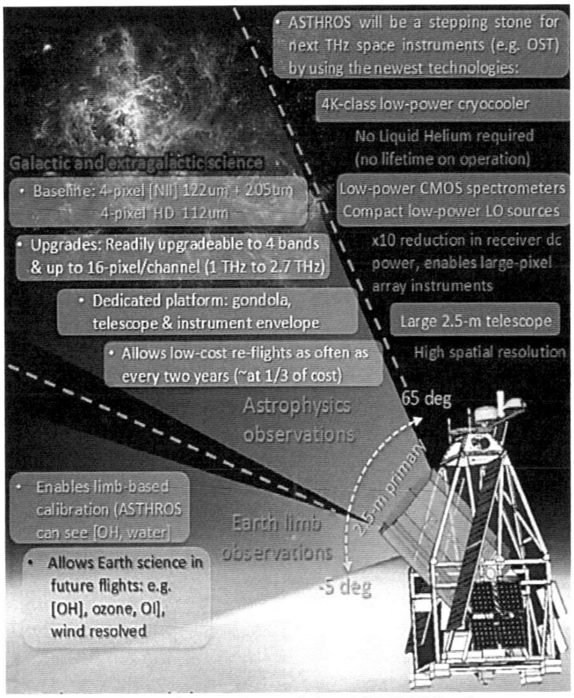

Fig. 1. General overview of the ASTHROS observatory capabilities.

We will present the overall mission science objectives and requirements, the overall payload architecture and initial subsystem designs and tests: telescope and optical subsystem, receiver subsystem, local oscillator subsystem, and gondola architecture.

[1] NASA Jet Propulsion Laboratory, California Institute of Technology, Pasadena, CA 91109, USA.
[2] Arizona State University, School of Earth and Space Exploration, Tempe, AZ 85287, USA
[3] Johns Hopkins University Applied Physics Laboratory, Laurel, MD 20723, USA.
[4] University of Miami, Coral Gables, FL 33124, USA.

A THz Space Mission to Probe the Trail of Water

Paul Goldsmith, Dariusz Lis, Youngmin Seo, Jon Kawamura, and Jose Siles
Jet Propulsion Laboratory, California Institute of Technology

We present a concept for a submillimeter spectroscopic mission to probe the trail of water from the interstellar medium to habitable planets. Water is an essential ingredient for life as we know it, and understanding how water is transported to planets in forming planetary systems is a fundamental question that we need to answer in order to understand how ocean worlds evolve throughout the universe. The water in habitable planets came from the protoplanetary disks out of which they formed, but a variety of obstacles make it impossible at the present time to determine observationally the distribution of water within such disks. Thus, we need to trace the water trail and understand the dense cores in which water can be formed, and how these regions evolve to become disks. And, we need to understand how the chemistry in disks evolves, and how material is transported within a disk, in order to determine the availability of water in different regions within them.

Observing gas-phase water in the local universe requires high spectral resolution because line widths may be less than 1 km/s in comets and dense cloud cores and a resolution of 0.1 km/s is compulsory. Thus, to study the water trail we need a heterodyne system, which is cryogenically cooled to minimize noise. To follow the water trail, we can consider a relatively large, ambient temperature telescope to maximize sensitivity and angular resolution, coupled to a multiband receiver covering key spectral lines of water and its isotopologues. A large survey of the submillimeter transitions of water vapor in conjunction with studies of water ice using JWST and SPHEREx will revolutionize our understanding of the role of water, its distribution, and key ISM processes. The required submm observations will not be possible with either the ESA/JAXA mission SPICA or the Astro2020 Decadal Study mission Origins Space Telescope (Origins) as developed by NASA, which in its current version includes a heterodyne instrument only as a possible up-scope option.

Three attributes for the receiver are: (1) to observe multiple bands simultaneously, (2) to have maximum sensitivity, and (3) to have spectral resolution ~0.1 km/s. The optical system first separates two linear polarizations, then separates the bands with frequency selective surfaces, and finally images beams coupling to the individual mixer feedhorns. We report on studies of a number of designs for a telescope of diameter between 2m and 7m. To observe lines between 500 GHz and 1200 GHz requires a surface accuracy of 13 microns *rms* or better. A number of advances have improved the ability to observe water in a wide range of sources including dense cores, protostellar disks, comets, and other minor objects in the solar system. These include 1) improved SIS receiver performance and design relative to what was available for Herschel/HIFI, 2) frequency-multiplied local oscillator chains that can now readily supply with significant tunability, and 3) CMOS ASIC digital spectrometers offer multi-GHz bandwidth per pixel with very low power consumption. These technical developments make a scientifically compelling Water Mission feasible and affordable.

The SAFARI far-IR Spectrometer for SPICA

Willem Jellema[1,2], Pieter Dieleman[1], and Peter Roelfsema[1,2] on behalf of the SAFARI consortium

The Space Infrared telescope for Cosmology and Astrophysics (SPICA) [1] is a joint space mission (Fig. 1) of the European Space Agency (ESA) and the Japanese Space Agency (JAXA), currently under study in ESA's M5 programme [2,3]. One of the proposed instruments is the SAFARI far-IR spectrometer providing low ($R \approx 300$) to medium (R up to 11000) resolution spectroscopy across 4 detector bands simultaneously covering the 34-230 μm wavelength band. Taking full advantage of a 2.5m cryogenic telescope, actively cooled to below 8K, SAFARI will offer unprecedented spectroscopic sensitivity closing the gap between JWST-MIRI and ALMA.

Fig. 1. The SPICA space observatory (JAXA/ESA/SPICA).

The instrument architecture of SAFARI is based on a cryogenic post-dispersed polarizing Martin-Puplett (MP) Fourier Transform Spectrometer (FTS) scheme, and 4 low-resolution grating spectrometer modules, employing horn-coupled TES detector arrays. The Transition-Edge Sensor (TES) detectors, cooled to about 50 mK by a Adiabatic Demagnetization Refrigerator (ADR) cryocooler, are combined with a Frequency Domain Multiplexed (FDM) readout system, and yield a detector system NEP of order 2×10^{-19} W/√Hz corresponding to a limiting line flux of about 5×10^{-20} W/m^2 (5σ-1hr).

In this paper we will provide an overview of the instrument design and its current state of development. We will present the optical design of the instrument, with a particular emphasis on the overall spatial multiplex and band division scheme, an 8x folded MP interferometer layout, and compact grating spectrometer modules. We also show recent results obtained in the key technology areas: the cryogenic FTS scan mechanism, the TES + FDM detector system and sub-Kelvin cryocooler. We conclude our paper by summarizing the expected performance metrics of the instrument.

References

[1] P.R. Roelfsema et al, "SPICA—A Large Cryogenic Infrared Space Telescope: Unveiling the Obscured Universe," *Publications of the Astronomical Society of Australia*, vol. 35, Aug.2018, 10.1017/pasa.2018.15

[2] https://spica-mission.org/spica-mission.org/version2/wp-content/uploads/2019/06/SPICA-M5-Proposal-V1.1.pdf

[3] https://spica-mission.org

[1] SRON Netherlands Institute for Space Research, P.O. Box 800, 9700 AV, Groningen, the Netherlands.

[2] Kapteyn Astronomical Institute, Faculty of Science and Engineering, University of Groningen, P.O. Box 800, 9700 AV, Groningen, the Netherlands

NOTES:

SUBMILLIMETER WAVE DIFFERENTIAL ABSORPTION RADAR FOR WATER VAPOR SOUNDING IN THE MARTIAL ATMOSPHERE

Omkar Pradhan, Ken Cooper, Leslie Tampari, Brian Drouin, Raquel Monje, Richard Roy, Jose Siles, Corey Cochrane

Abstract

In this study we report on the current state-of-progress of the Water Sounding Short-range Radar (WASSR) project. WASSR is a frequency modulated continuous wave (FMCW) differential absorption radar (DAR) instrument and will facilitate estimation of near-surface water vapor profile on Mars. The scope of the work presented here includes a discussion of (i) the science background and motivation for profiling water vapor near the Martian surface and (ii) submillimeter wave radar system implementation and characterization.

1. Science Background and Motivation

Improved atmospheric modelling and incorporation of additional weather and climate processes into the models has resulted in a renewed interest in even the small amount of water vapor found in the Martian atmosphere. For instance the incorporation of the tenuous Martian clouds and their heating and cooling effects in Mars climate models showed a surprisingly large effect on the atmospheric processes [1], [2], [3]. Furthermore the presence of perchlorate, an extremely hygroscopic salt, as identified first by the wet chemistry experiment onboard the Phoenix lander [3] points to interesting water vapor-surface interactions that were previously thought to be minor and unimportant. The near-surface water vapor profile estimation made possible with the WASSR instrument can thus provide critical measurements that facilitate better understanding of atmospheric water vapor dynamics as well as it's interaction with the Martian surface.

2. System Implementation and Characterization

The WASSR system is currently being assembled for bench-top testing and algorithm implementation. Specifically, the RF back-end components have been procured and a brass-board setup has been constructed for FMCW signal generation and characterization. SMMW components such as tripler and integrated sub-harmonic mixer and horn, both developed internally at JPL, are being characterized for conversion efficiency and noise figure. Results of these characterizations along with achievable water vapor profiling sensitivity will be presented and discussed at the ISSTT 2020 meeting. Furthermore, algorithm validation using a gas-cell spectrometer-type setup is planned and will be presented as well. This setup is critical for a proof-of-concept demonstration of the applicability of the DAR technique for water vapor profiling at Mars-like meteorological conditions.

References

[1] J.-B. Madeleine, F. Forget, E. Milour, L. Montabone, and J. WOlff, "Revisiting the radiative impact of dust on Mars using the LMD Global Climate Model," *J. Geophys. Res.*, vol. E11, no. 116, 2011.

[2] R. A. Urata and O. B. Toon, "Simulations of the Martian Hydrologic Cycle with at General Circulation Model: Implications for the Ancient Martian Climate," *Icarus2*, vol. 1, no. 226, pp. 229–250, 2013.

[3] M. H. Hecht *et al.*, "Detection of perchlorate and the soluble chemistry of martian soil at the phoenix lander site," *Science (80-.).*, vol. 325, no. 5936, pp. 64–67, Jul. 2009.

Millimetron Space Observatory Mission Development

A. Baryshev[1], A. Smirnov[2], E. Golubev[2], M. Arkhipov[2], E. Filina[2], V. Pyshnov[2], N. Myshonkova[2], S. Fedorchuk[2], I. Vinogradov[2], Th. de Graauw[2], S. Likhachev[2]

Millimetron Space Observatory (MSO) is mission of Russian Space Agency devoted to astrophysical observations in the far infrared wavelength regime. It is built around a space telescope which is based on deployable 10 m diameter main reflector. The observatory is planned to be launched in the timeframe of 2029. The main working orbit of MSO will be a halo around the Sun-Earth Lagranian point L2. This orbit has an excellent thermal environment and this allows to actively cool main dish to cryogenic temperatures of below 10 K. Low antenna temperature allows Millimetron mission to achieve a very low background limited sensitivity for a direct detector based spectrometer type instruments. MSO will work in two major regimes: Space-Earth interferometer and single dish observatory modes. For the Space-Earth very large baseline interferometer (SE-VLBI), Millimetron will work together with Earth ground telescopes such as Event Hirisont Telescope (EHT) and other millimeter submillimeter facilities to achieve extremely high spatial resolution images of nearby objects. In order to improve a u-v coverage of Millimetron SE-VLBI, after operating at the L2 orbit, the satellite will be pushed on to highly elongated elliptical orbit with perigee and apogee of 10000 km and 300000 km respectively. Orbit period will be 10 days which, in combination with EHT ground based telescopes allows to build dynamic images of event horizon shadow of massive black holes in the center of our galaxy and other large galaxies such as M87.

The Millimetron antenna has central non-deployable mirror of 3 m diameter which consists of 24 panels and 24 deployable petals each of which has 3 surface panels. Panels of the telescope are made from carbon fiber reinforced composite material which is formed on top of the highly accurate and highly dimensionally stable Astrosital glass mould. This technique allows to achieve high panel accuracy of order of 4.7 micron RMS surface deviation from an ideal parabola, while keeping the specific mass of the antenna structure very small 8.5 kg/m^2.

The antenna and instrument compartment is protected from Solar infrared radiation by a 5 layer deployable screens which allow to minimize the parasitic thermal load on the antenna to the levels of below 100mW at 4 K. This makes it feasible to use space qualified closed cycled refrigerators to actively cool both the main dish and the instrument compartment.

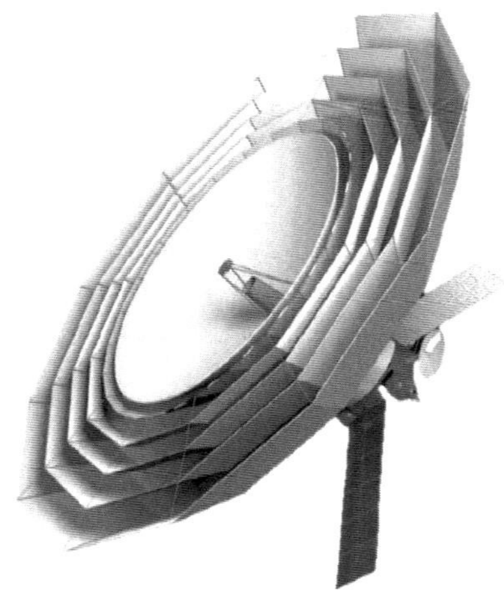

Fig. 1. Millimetron Space Observatory in deployed configuration .

Antenna panels are mounted on the support structure by using actuators. Laser based absolute metrology system will be used for the in-flight panel position measurements and alignment.

Millimetron instrumentation will include both heterodyne type to support VLBI mode and high resolution spectroscopy and both grating and Fourier transform type direct detector based spectrometers. Millimetron instruments will cover from 40 GHz up to 4.7 THz frequency range.

In this contribution we will report current status of Millimetron mission development and design solutions and achieved parameters for main subsystems such as antenna structure, antenna alignment system, instrument content, thermal system and optics. We also summarize key science cases as well as expected on-board instrumentation parameters and requirements.

[1] Kapteyn Astronomical Institute, University of Groningen, Landleven 12, 9747 AD, Groningen, The Netherlands..

[2] Astro Space Center of P.N. Lebedev Physical Institute, 84/32 Profsoyuznaya str., Moscow, Russia.

Submillimeter Wave Power Generation Using InP HEMT Technology

W.R. Deal[1], Kevin Leong[1], Caitlyn Cooke[1] and Xiaobing Mei[1]

Over the last decade, Indium Phosphide High Electron Mobility Transistors (InP HEMT) technology has pushed to increasingly higher operating frequencies. Significantly, room temperature amplification has been demonstrated to 1 THz, and transistor maximum frequency of oscillation (f_{MAX}) has been reported to 1.5 THz [1]. These terahertz transistors are fabricated in a monolithic integrated circuit technology. Results for amplification have been measured both on-wafer, and in split block packages.

These increases in operating frequency have come through transistor gate scaling and optimization. Described in [1], the highest reported InP HEMT node uses a 25 nm gate length. Earlier generations (100 nm) of this technology are highly mature, and have been used in space applications for many years.

To date, most application of this technology has focused on realizing sensitive receivers. Significant efforts have been made to develop sensitive receivers using this technology [2], [3]. InP HEMT receivers have demonstrated noise temperatures of 2,000-2,500 K at 670 GHz [2] and 4,300 K at 850 GHz [3]. Note that these are LNA based front-ends, and noise temperatures are analogous to mixer single-sideband noise temperature.

Less work has been reported on output power or power generation efficiencies at these frequencies. We will describe results from 100-1,000 GHz. Power benchmarks will be reported, along with DC efficiencies. Recent results on 330-410 GHz solid state power amplifiers will be reported as well. Output powers of 8-10 mW are obtained from 330-380 GHz, and 6-9 mW from 360-410 GHz from two different packaged MMIC designs. These results are obtained from a single packaged MMIC.

We also provide a summary of InP HEMT multiplier chain results. It will be shown that InP HEMT multiplier chains provide significant benefits for DC efficiency at submillimeter wave frequencies because the majority of power is consumed by the power amplifier at the output of the InP HEMT multiplier chain. These InP HEMT multiplier chains provide sufficient power to serve as the local oscillator for mixers.

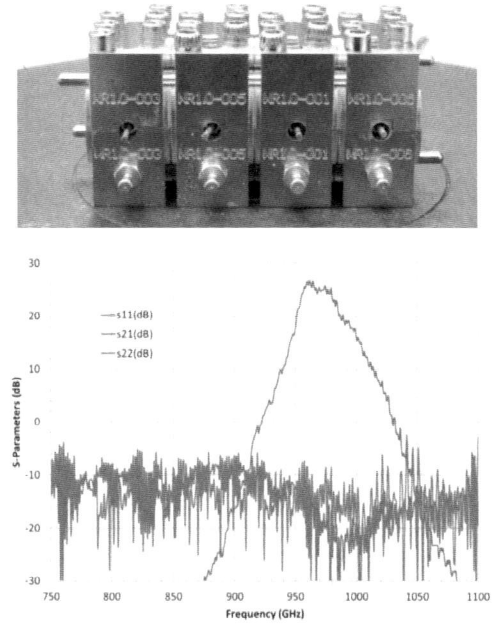

Fig. 1. 1 THz InP HEMT amplifier, including packaged amplifiers (Top) and measured small signal gain response (bottom)

References

[1] X.B. Mei et al., "First demonstration of amplification at 1 THz using 25-nm InP high electron mobility transistor process," IEEE Electron Device Lett., vol. 36, no. 4, pp. 327-329, Mar. 2015.

[2] W.R. Deal et al, "A low-power 670-GHz InP HEMT receiver" IEEE Trans. THz Sci. Technol, vol 6, no. 6, pp. 862-864, Nov. 2016.

[3] K.M.K.H Leong et al., "850 GHz receiver and transmitter front-ends using InP HEMT", " IEEE Trans. THz Sci. Technol, vol 7, no. 4, pp. 466-475, Jun. 2017.

[1] Northrop Grumman Space Systems, Redondo Beach, CA 90272 USA.

Technology Roadmap for the HEterodyne Receiver for Origins (HERO)

M. C. Wiedner[1], A. Baryshev[2], V. Belitsky[3], V. Desmaris[3], A. DiGiorgio[4], J,-D. Gallego[5], M. Gerin[1], P. Goldsmith[6], F. Helmich[7], W. Jellema[7], A. Laurens[8], I. Mehd[6], C. Risacher[9], J. Bardin[10], S. Cherednichenko[11], J.R. Gao[12], P. Grimes[13], C. Groppi[14], E. Tong[13], S. Weinreb[15], L. Zeng[13], HERO teams and the Origins mission concept study team

The Origins Space Telescope is one of four large missions studied by NASA, which have been submitted to the 2020 Decadal Survey on Astronomy and Astrophysics.

The Origins Space Telescope is a mid to far-IR satellite with a cooled 5.9m primary mirror. Origins has three bolometer arrays as the baseline instruments and one upgrade instrument, the HEterodyne Receiver for Origins (HERO). HERO [1] is designed to study the trail of water from the ISM to protoplanetary disks. It covers a continuous wavelength range of 486 to 2700 GHz in only 4 bands. Each band has two 9-pixel focal plane arrays, one for each polarization. Hero can carry out dual-polarization and dual-frequency observations. We aim for DSB receiver temperatures around 50 K for the 486 to 756 GHz band, 100K for the 756 - 1188 GHz band, 200K for the 1188 - 1782 GHz band and 300K for the 1782 to 2700 GHz band. HERO has at least 6GHz IF bandwidth (goal 8 GHz).

[1] Sorbonne Université, Observatoire de Paris, Université PSL, CNRS, LERMA, F-75014, Paris, France

[2] Kapteyn Astronomical Institute, University of Groningen, Postbus 800, 9700 AV, Groningen, The

[3] Group for Advanced Receiver Development (GARD), Department of Space, Earth and Environment, Chalmers University of Technology, 41296, Gothenburg, Sweden.

[4] Istituto Nazionale di Astrofisica-Istituto di Astrofisica e Planetologia Spaziali, Via Fosso del Cavaliere 100, I-00133 Roma, Italy.

[5] Centro Astronómico de Yebes, Observatorio Astronómico Nacional, Apdo. 148, 19080 Guadalajara, Spain,

[6] Jet Propulsion Laboratory, which is operated for NASA by the California Institute of Technology.

[7] SRON Netherlands Institute for Space Research, Groningen, The Netherlands and Kapteyn Astronomical Institute, University of Groningen, The Netherlands.

[8] Centre National d'Études Spatiales, 18 Avenue Edouard Belin, 31400 Toulouse, France.

[9] Max-Planck-Institut für Radioastronomie, Auf dem Hügel 69, 53121, Bonn, Germany and is now at IRAM, 300 rue de la Piscine, 38406, St. Martin d'Hères, France,

[10] University of Massachusetts Amherst, Amherst, USA,

[11] Department of Microtechnology and Nanoscience, Chalmers University of Technology, 41296, Gothenburg, Sweden,

[12] SRON Netherlands Institute for Space Research and Delft University of Technology,

[13] Harvard Smithsonian Center for Astrophysics, Cambridge, USA,

[14] Arizona State University, Tempe, USA,

[14] Caltech, Pasadena, USA.

Sub-system	Component	HIFI	HERO
Local Oscillator	Synthesizer Technology	YIG-based Ka-Band+Active Frequency Multipliers + GaAs W-band power amps	CMOS or YIG-based W-band synthesizer + GaN. amps
	Multiplied LO	Cascaded GaAs frequency multipliers	Cascaded Multipl. + On-chip. Power Combining +. 3D integ.
	Pixels per array	1	9
	DC power/pixel	25 W	~ 1.5 W
	Fractional Bandwidth	~12 %	~45 %
Cryogenic Detectors	Mixer Technology	SIS, HEB	SIS, HEB
	SIS sensitivity	2 - 6 hv/k	2 hv/k
	HEB sensitivity	13-18 hv/k	3 hv/k
	LNA Technology	InP HEMT	Low-power SiGe HBT
	Number of pixels	1	2x9
	DC power/pixel	10 mW	0.5 mW
	Mixer. Assembly	Quasi-optical	Waveguide
Backend	IF Processing	GaAs HEMT amplifiers	
	Spectrometer Tech.	FFT FPG	CMOS based SoC
	DC Power/pixel	10 W	2W
	IF Bandwidth	1.5 GHz	6 GHz goal 8GHz
Total DC power per pixel		35 W	3.5 W

Fig. 1. HERO builds on the successful HIFI/Herschel mission, but carries it to new dimensions by proposing the first heterodyne array receiver for a satellite.

On a satellite, resources are scarce and we require a drastic reduction in cooling power and electrical power with improved instrument performance (see table above). HERO builds on promising recent developments, and we propose a dedicated technology roadmap to reach TRL 5 in 2025 for all components.

REFERENCES

[1] M. C. Wiedner et al., "*Proposed Heterodyne Receiver for the Origins Space Telescope* " *IEEE Trans. On THz Science and technology*, vol 8, issue 6, pp. 558-571.

NOTES:

High-Sensitivity Terahertz Detection Module (HSTDM) onboard China's Space Station

Sheng-Cai Shi[1,2] with HSTDM team

The Terahertz (THz) regime is of particular interest in astronomy as ones can observe radiation from early distant, cold, and dusty objects, which are mostly invisible in the optical/NIR regime. It, however, is still a frequency band to be fully explored as water vapor renders the terrestrial atmosphere opaque over nearly all of the Earth's surface. China is going to launch a space station around 2022, providing a good opportunity for the development of China's space astronomy. And there will be a 2-m optical telescope flying simultaneously with the space station in common orbit. Five astronomical instruments including high-sensitivity terahertz detection module (HSTDM) have been selected onboard the 2-m optical telescope facility. The HSTDM, focusing mainly on the observations of neutral carbon (CI) and other molecular lines for the understanding of star formation, is mainly a 500-GHz niobium-nitride (NbN) superconducting-tunnel-junction (SIS) heterodyne receiver [1] operated at a temperature of 8-10 K, which is provided by a closed-cycle 2-stage pulse-tube cryocooler [2]. In addition, the possibility of adding a THz Schottky receiver onto HSTDM is taken into account. Detailed designs and testing results will be presented.

[1] Purple Mountain Observatory, Chinese Academy of Sciences, Nanjing 210033, China.

[2] Key Lab of Radio Astronomy, , Chinese Academy of Sciences, Nanjing 210033, China.

References

[1] Jing Li et al., "Low-noise 0.5 THz all-NbN superconductor-insulator-superconductor mixer for submillimeter wave astronomy," *Appl. Phys. Lett.*, vol. 92, pp. 222504-1/3, 2008, 10.1063/1.2940235.

[2] Jia Quan et al., "4 K high frequency pulse tube cryocooler used for terahertz space application," *Chin. Sci. Bull.*, vol. 59, pp. 3490-3494, 2014, 10.1007/s11434-014-0457-5.

NOTES:

A Single-Sideband 530-600GHz Integrated Receiver Utilizing Tunable Waveguide Filters

Theodore Reck[1], Daniel Koller[1], Jeffrey Hesler[1], and Eric Bryerton[1].

Sideband separating receivers radically simplify analysis of spectral regions where many species exist, such as the 530-600 GHz band when studying planetary atmospheres. Sideband-separating mixers use quadrature combining of two mixers to down-convert both sidebands but require careful balancing of the two mixer paths. A simpler approach is possible if a fixed LO frequency can be used. This allows filtering of the image-band before a double-sideband mixer. In addition to reducing circuit complexity, this approach provides higher sideband rejection than typically achieved with sideband separating mixers.

Fig 1 (top) shows the system diagram. The integrated receiver combines a 281GHz LO chain with a multiplication factor of 27 with two sub-harmonic mixers. IF filters follow each mixer. Both sidebands are down-converted by splitting the incoming signal and then filtering each path. The LO is located at 562GHz and down-converts 531-557GHz and 568-600GHz, providing an 11GHz band where the filters can cross-over. Secondary down-conversion is used to place the species of interest into the spectrometer's bandwidth. To set the noise temperature of the system, 520-600GHz LNAs will precede the splitter.

This approach requires very precise location of the passband edges. Machining tolerances makes realization of the desired filters' passbands unlikely, so a tunable waveguide filter approach is utilized. It has been found that waveguide filters can be effectively tuned by adjusting the gap in the E-plane split waveguide block [1]. By controlling the gap in the filter block with 5um thick shims, the filters can be aligned to within +/- 3 GHz, without significant change in the passband width or insertion loss.

Fig. 1 (middle) shows the discrete version of this system, where the tunable filter and splitter have been separated from the integrated receiver to verify this tunable filter approach. The graph shows the conversion gain of each mixer path, which includes the loss of the splitter, filter and mixer, and the gain of an IF amplifier.

To further reduce losses, a fully integrated version of the system that combines the receiver with the filter and splitter has been developed and results of this device will be presented at the conference.

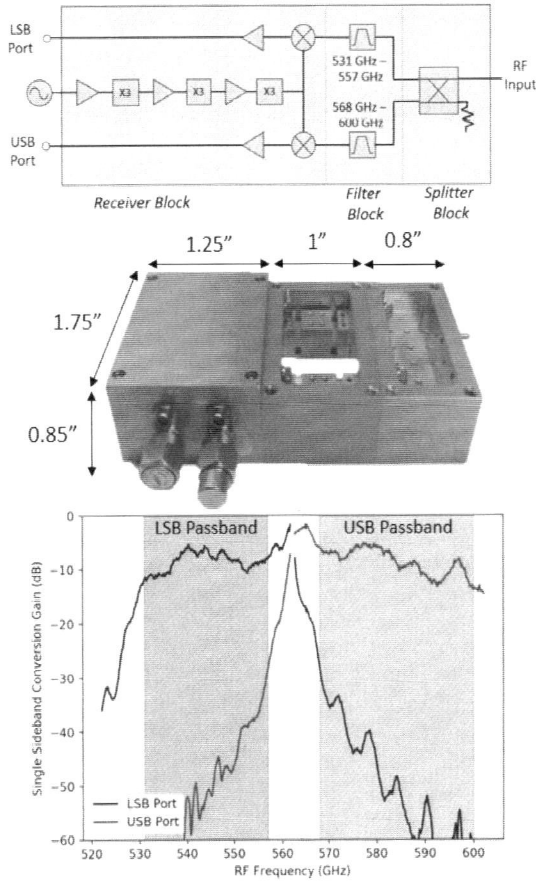

Fig. 1. (Top) The block diagram of the 530-600GHz receiver. (Middle) The discrete 530-600 GHz receiver system in three blocks. From left to right. Integrated receiver, Tunable Filter, Splitter. (Bottom) is the conversion loss of the full system showing less than -10 dB conversion loss and over 30 dB of sideband rejection across both upper and lower passbands.

References

[1] D. Koller, E. W. Bryerton and J. L. Hesler, "WM380 (675–700 GHz) Bandpass Filters in Milled, Split-Block Construction," in IEEE Transactions on Terahertz Science and Technology, vol. 8, no. 6, pp. 630-637, Nov. 2018.

[1] Virginia Diodes Inc., Charlottesville, VA 22902 USA.

180 GHz CMOS Pulsed Transmitter and Heterodyne Receiver Pair for in-situ Chemical Detections

Deacon J. Nemchick[1], Brian J. Drouin[1], Adrian J. Tang[2,3], Yanghyo (Rod) Kim[3], Maria Alonso[2], M.C. Frank Chang[3]

Recent[1-3] work has shown the coherent radiation generated by high-speed phase lock loops (PLLs) embedded in CMOS circuitry can be leveraged for use in a variety of remote and in-situ sensing instrumentation. PLL outputs can be paired with other on chip elements (e.g., down-conversion mixers, power amplifiers, frequency multipliers, pulse modulators) to produce versatile transmitter or receiver devices having a footprint of several square millimeters that consume <500mW input power. A previous joint JPL/UCLA collaboration to develop a miniaturized 90-100 GHz cavity enhanced pulsed-emission spectrometer for *in-situ* sensing deploys such custom CMOS source/detection electronics. This sensor, where all source and detection electronics are housed on a single (100 cm^2) printed circuit board, has been shown to be sensitive to trace amounts (1-100 μTorr) of test gasses with short (1 second) integration times.

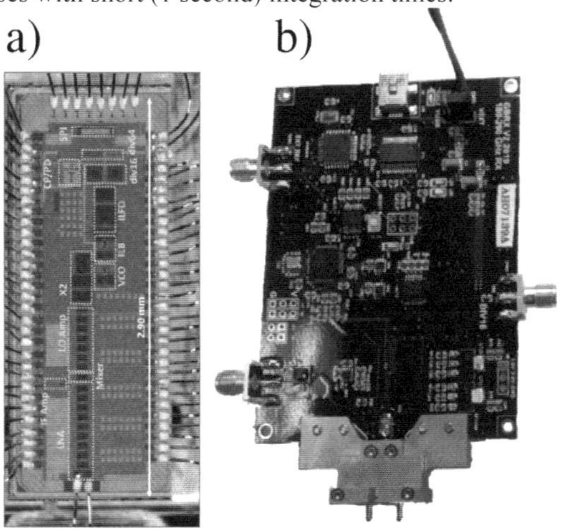

Fig. 1. Pictorial representation of custom designed 180 GHz CMOS heterodyne receiver (a). This custom integrated circuit chip is mounted on a printed circuit board (b) that houses all support electronics. A standard WR5 waveguide flange is used for interfacing with standard mm-wave hardware for laboratory characterization.

This talk will focus on the next generation of instrumentation currently under development that aims to realize a dual band system capable of detecting both H_2O (at 183.310 GHz) and HDO (at 80.3578 GHz) to allow for localized determinations of H/D ratios. A full system description will be provided along with an introduction to the pulsed-emission detection scheme leveraged by the miniaturized spectrometer. The performance properties of the 180 GHz pulsed transmitter and heterodyne receiver chip set (*viz*. Fig. 1) will be discussed along with details of how generated radiation is coupled into (and out of) a resonant optical cavity. These results will be accompanied by demonstrative examples of water vapor detections and future plans to integrate a lower frequency band for dual H_2O/HDO detections.

References

[1] D. J. Nemchick, B. J. Drouin, A. J. Tang, Y. Kim, M.-C. F. Chang, Sub-Doppler Spectroscopy with a CMOS Transmitter, IEEE Trans. THz Sci. Technol., 8, 121-126 (2018)

[2] D. J. Nemchick, B. J. Drouin, M. J. Cich, T. Crawford, A. J. Tang, Y. Kim, T. J. Reck, E. T. Schlecht M.-C. F. Chang, G. Virbila, A 90- 102 GHz CMOS Based Pulsed Fourier Transform Spectrometer: New Approaches for *in situ* Chemical Detection and Millimeter-Wave Cavity-Based Molecular Spectroscopy, Rev. Sci. Inst., 89, 073109:1-12 (2018)

[3] Y. Kim, Z. Yan, T. Reck, D. Nemchick, G. Chattopadhyay, B. Drouin, M.-C. F. Chang, A. Tang, A 183 GHz InP/CMOS-Hybrid Heterodyne-Spectrometer for Spaceborne Atmospheric Remote Sensing, IEEE Trans. THz Sci. Technol., 9, 313-334 (2019)

[1] Laboratory Studies and Atmospheric Observation, Jet Propulsion Laboratory, California Institute of Technology, Pasadena, CA, 91109

[2] Submillimeter wave Technology, Laboratory Studies and Atmospheric Observation, Jet Propulsion Laboratory, California Institute of Technology, Pasadena, CA, 91109

[3] Electrical Engineering, University of California – Los Angeles, Los Angeles, CA, 90095

NOTES:

Integrated Silicon Platform for Co-planar Design of Vertically Stacked 2.06 THz Mixer Module

Christine P. Chen, Cecile Jung-Kubiak, Robert Lin, Darren Hayton, Jose Siles, Joseph Lee, Alex Peralta, Alain Maestrini, and Imran Mehdi

***Abstract*— A hybrid multiplier chain with a Si-micromachined stack housing the 2.06-THz sub-harmonic Schottky-diode based mixer is developed for the purpose of THz sensing for NASA in space. Low-noise temperature measurements are performed using this hybrid local-oscillator (LO) chain. The capability to have exact dimensionality becomes essential as frequency scales due to the complexity arising in assembly and finer tolerances. Total process yields can be improved through Si fabrication by verifying achievement of specified tolerances lithographically and compatibility to rapid process changes and design fine-tuning. The integrated Si platform is detailed, whereby signal I/O and alignment modalities enable low-noise measurements.**

***Index Terms*—Silicon, Space technology, Submillimeter wave propagation, System integration, Wafer scale integration**

I. Introduction

UNDERSTANDING the atmospheric dynamics in Earth's upper atmosphere is an area of focus for NASA's earth science directorate [1]. Schottky diode mixers operating at several THz have historically been a stable method of passive sensing in space, with prior scientific mission deployments [2].

Figure 1. a) Device mounted in the Si block displayed in b). The beam lead of the diode is aligned to the center of the waveguide, and achieves precise alignment for complete coupling of the mode [3].

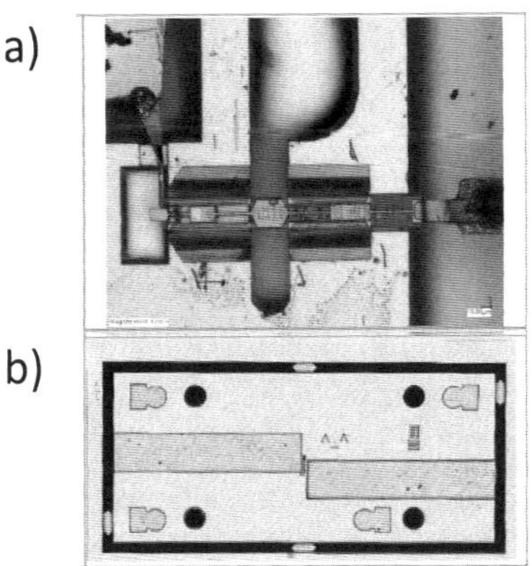

The capability to leverage silicon (Si) fabrication and integrate the local oscillator (LO) and intermediate frequency (IF) signal onto a compact Si micro-machined package has the potential to introduce new features and design paradigms.

A heterodyne receiver front end is being developed to perform measurements at 2.06 THz, where a neutral oxygen [OI] line exists. By measuring this atmospheric feature, thermos-spheric models can be created for understanding space weather and its impact on earth climate.

In this paper, we describe the process for fabricating and utilizing this three-dimensional Si stack receiver front-end. Silicon micromachining provides the necessary accuracy for integration at the several THz regime. Particularly, processing and etch recipes have demonstrated smooth surfaces of < 1 μm surface roughness. We keep in consideration pre-compensating for minute offsets, matching mechanical strain on all interfaces, and including passive alignment verification in the design. As compared to metal blocks, computer numerical controlled (CNC) milling can produce slight burrs in the interfaces resulting in misalignment and resultant strain between two pieces of components. Si integration can also make possible arrays of pixel transceivers through compact subsystems, drastically minimizing unit area.

II. Background

Silicon micromachining is used to create the fine waveguide features in the three-dimensional stack. The fabrication process consists of two main steps. In order to etch straight side walls, multiple SiO_2 masks are used. This type of process allows for thicker structures and multiple-levels for different feature heights [4]. The recipe used in this particular paper was first developed and described in [5] at Jet Propulsion Laboratory and adapted to be used for this work.

A three-step process for deep-reactive ion etching is performed. Negative photoresist is lithographically patterned on the 4-inch Si wafer. With the pre-determined selectivity of the process, inductively coupled plasma (ICP) etch is initially performed on the SiO_2. This step allows for finer precision over final target thicknesses. Then, deep-reactive ion etching (DRIE) is done, creating the desired depths of the Si waveguide. Waveguide depths are measured using a profilometer, in order ensure accuracy of process, which will be discussed further in the paper.

III. SILICON INTEGRATION

The Si module, encompassing only a miniature area of 7.5 mm x 17.5 mm x .35 mm, is assembled in a hybrid package to interface to the rest of the metal chain. Within the module is housed a biased diode structure, as shown in Figure 1a, and the intermodulation frequency (IF) line. Two wafers of the design were fabricated, with fourteen top and bottom Si sets per wafer.

Fig. 2. The waveguide structure and its critical dimensions, mostly etch depths, are measured and their deviation from design are shown in this diagram. These values are well within tolerance of retaining optimal performance.

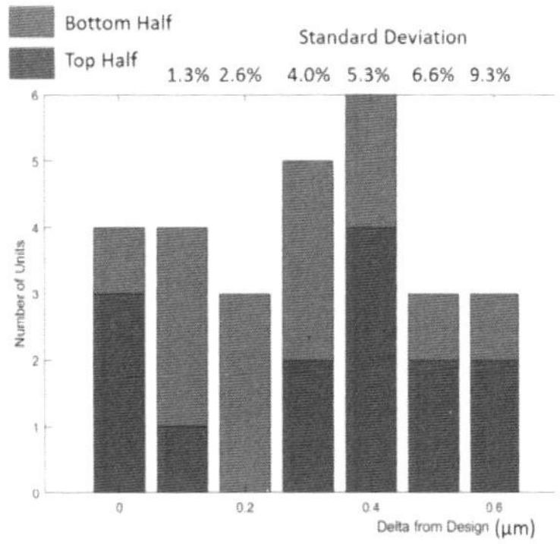

As discussed earlier, multi-step DRIE allows for Si micromachining of multiple depths with precision and smoothness. Other features shown in the waveguide picture in Figure 1b are designed to be within their required depths, as well as match complementary etch depths.

The exact dimensions are met with accuracy by reduced etches and checks with the calibrated Dektak profilometer. Figure 2 illustrates the delta from designed waveguide depth measured across wafer.

IV. EXPERIMENTAL RESULTS

Experimental results are demonstrated using the LO chain, represented by Figure 3, consisting of a National Instruments synthesizer source amplified by a Cernex power amplifier to drive the 110 GHz tripler and a Gallium Nitride (GaN) amplifier. From here, the signal is launched into the 340 GHz tripler, followed by a 1 THz doubler providing 1 milliwatt (mW) of power at various frequency points into the 2.06 THz subharmonic mixer.

In Silicon, with RF pumping at 1.987 THz, the mounted diode is operational with > 0.150 mA current at 1.3 V reverse bias. For the balanced diode, with series resistance in the 100 to 170 Ω range, noise temperature measurements are predicted to be in a similar to lower range as the demonstration in the metal waveguide enclosure.

Fig. 3. Block diagram of experimental setup for room temperature measurements b) Picture of setup during noise temperature measurement, with the 2.06 THz mixer at the rightmost end of the local oscillator chain.

a)

b)

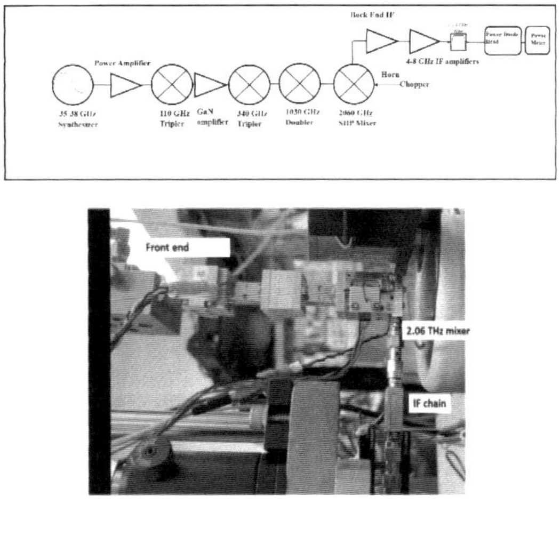

V. CONCLUSION

NASA has as one of its central goals to model thermo-spheric dynamics, and this work provides the necessary technology to enable the measurement of this spectral characteristic.

The advantage of utilizing this Si process for space technology is twofold. Beyond rapid prototyping, the designed stack is ready for flight demonstrations and extensible to array form. This particular interface will be directly applied to the detection of the OI line, which occurs at 2.06 THz frequency.

Prior measurements at > 2 THz have not been at room temperature, which is targeted with this heterodyne detection scheme [6] for reduced power consumption.

ACKNOWLEDGMENT

The authors are at the NASA/Jet Propulsion Laboratory, California Institute of Technology, Pasadena, CA, 91109, with C. P. Chen under the NPP program, which is administered by the Universities Space Research Association under contract with NASA.

REFERENCES

[1] "Atmospheric Composition" https://science.nasa.gov/earth-science/programs/research-analysis/atmospheric-composition, 2018.

[2] I. Mehdi, "THz Instruments for Space Exploration," *Proceedings of 2017 Asia Pacific Microwave Conference*, 2017.

[3] D. Pozar, Microwave Engineering, Wiley Global Education, 2011.

[4] Y. Mita, *et al.* "Embedded-Mask-Methods for mm-scale multi-layer vertical/slanted Si structures," *Proceedings IEEE Thirteenth Annual International Conference on Micro Electro Mechanical Systems*, 2000.

[5] C. Jung-Kubiak, *et al.* "A Multistep DRIE Process for Complex Terahertz Waveguide Components," *IEEE Transactions on Terahertz Science and Technology*, vol. 6, no. 5, Sept., 2016.

[6] Siegel, Peter, *et al.* "2.5-THz GaAs Monolithic Membrane-Diode Mixer," IEEE Transactions on Microwave Theory and Techniques, vol. 47, no. 5, May, 1999.

Heterodyne receiving with frequency combs: towards simultaneous ultrabroadband spectroscopy

David Burghoff[1]

A fundamental problem is detecting the spectrum of a remote source, and heterodyne spectroscopy is a powerful technique for high-resolution measurements. However, performing these measurements over a wide range requires a widely tunable local oscillator. In THz quantum cascade lasers, this typically requires moving parts [1] and requires nontrivial amplitude and frequency calibration.

One potential alternative is the use of optical frequency combs as LOs. Frequency combs are light sources whose lines are perfectly evenly spaced [2], and in THz QCLs have achieved bandwidths exceeding 1 THz. For local measurements, the dual-comb spectroscopy technique—a multiheterodyne technique in which two different combs are beat together [3]—has become essential. However, measuring the spectrum of a *remote* source with a comb is more challenging. Though one can beat a comb LO with a signal, the resulting beating is ambiguous in the sense that the resulting beatings cannot be attributed to a specific line.

We will show that a new kind of dual-comb spectrometer, in combination with a fast analyzer, can be used to unambiguously determine the spectrum of a remote source over the comb's bandwidth. By beating the signal with two different combs, the original signal can be unambiguously retrieved, merely by computing the Fourier transform of the correlation between the spectrum of the two signals. (Alternatively, one comb can be used if a delay element is introduced.) This approach is suitable for any remote signal, even incoherent ones. In addition, the presence of multiple LOs allows for the double-sideband ambiguity to be resolved even when a double-sideband mixer is used.

Figure 1 simulates this process for a pair of LO combs spanning 4 to 5 THz at 10 GHz increments and a signal with lines at 4.211 THz, 4.558 THz, 4.783 THz, and 4.883 THz. Fig. 1a shows the beating with the first comb alone, showing components at 1, 2, and 3 GHz. Though beatings are present, which comb line they correspond to is ambiguous. However, by beating the signal with the second comb and calculating the Fourier transform of the correlations between the two spectra, one can determine which comb line each beating belongs to. These are shown below for the comb lines at 4.21 THz, 4.56 THz, 4.78 THz, and 4.88 THz. The true signals (shown in red) are reproduced, showing the four components at offset frequencies of 1 GHz, -2 GHz, 3 GHz, and 3 GHz, respectively. Note that both negative frequencies and overlapping frequencies in the IF were correctly reproduced, even though a double-sideband mixer was used.

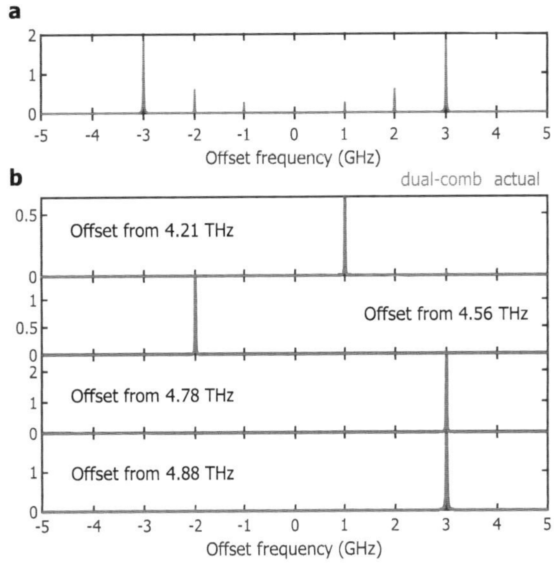

Fig. 1. a. Beating of a signal with four lines with a 4-5 THz comb. All beatings are present, but the result is ambiguous. b. Signal retrieved by a Fourier transform of the frequency-domain correlation at four comb frequencies. The signal is reproduced.

In principle, this system can be compact, monolithic, and self-calibrated—both in amplitude and frequency. We will also show that under certain conditions (the case of a signal whose IF components do not overlap), such systems can inherit the sensitivity of an equivalent system with a tunable LO. Thus, this scheme can act as an alternative route for ultrasensitive, ultrabroadband, high-resolution spectroscopy.

References

[1] C. A. Curwen, J. L. Reno, and B. S. Williams, "Broadband continuous single-mode tuning of a short-cavity quantum-cascade VECSEL," *Nat. Photonics*, pp. 1–5, Sep. 2019.

[2] D. Burghoff *et al.*, "Terahertz laser frequency combs," *Nat. Photonics*, vol. 8, no. 6, pp. 462–467, Jun. 2014.

[3] D. W. V. der Weide, J. Murakowski, and F. Keilmann, "Gas-absorption spectroscopy with electronic terahertz techniques," *IEEE Trans. Microw. Theory Tech.*, vol. 48, no. 4, pp. 740–743, Apr. 2000.

[1] Department of Electrical Engineering, University of Notre Dame, Notre Dame, IN 46656.

Wide-band Superconducting Parametric Amplifiers for Millimeter-Wave Instruments

Peter K. Day[1], Nikita Klimovich[2], Byeong Ho Eom[1], and Henry G. Leduc[1]

Recently, superconducting parametric amplifiers have been developed that are able to provide gain over a very wide instantaneous bandwidth, extending up to an octave at microwave frequencies. These amplifiers overcome the limited bandwidth of previous implementations of paramps that relied on nonlinear components embedded in resonant circuits while maintaining extremely low noise behavior that is limited only by quantum mechanical constraints. The new designs are based on a traveling wave architecture where the nonlinearity of the circuit is distributed along the length of a non-resonant transmission line structure. An example of this sort of device makes use of the nonlinear kinetic inductance of a thin film superconducting transmission line [1].

Kinetic inductance traveling wave parametric amplifiers may be interesting for millimeter-wave instruments either as very low noise front-end amplifiers or as microwave band IF amplifiers. The use of a paramp front end amplifier could allow for an improvement in system noise temperature beyond what has been achieved with transistor amplifiers or SIS mixers. We have observed gain at around 90 GHz in a waveguide-coupled paramp, demonstrating the potential of these devices at millimeter wavelengths.

We will also discuss results on microwave band paramps and their possible use as IF amplifiers in heterodyne receiver systems. These paramps dissipate only a few microwatts of power, compared to a few mW transistor LNA amplifiers, making them interesting for arrays. The added noise is measured to be close to the quantum limit over about an octave bandwidth.

References

[1] Ho Eom, B., Day, P., LeDuc, H. *et al.* A wideband, low-noise superconducting amplifier with high dynamic range. *Nature Phys* **8,** 623–627 (2012) doi:10.1038/nphys2356

1 Jet Propulsion Laboratory, Pasadena, CA 91107 USA.
2 California Institute of Technology, Pasadena, CA 91125 USA.

Mixing with Y-Ba-Cu-O Josephson Junctions Fabricated with Focused Helium Ion Beam Irradiation

Anthony Cortez[1], Ethan Y. Cho[2], Hao Li[2], Daniel Cunnane[3], Boris Karasik[3], and Shane A. Cybart[2]

Current state-of-the-art mixers with the best reported sensitivities are superconductor-insulator-superconductor (SIS) receivers and hot-electron bolometers (HEB) [1]. Both have proven to be successful in the practicality of space flight missions. However, SIS and HEB mixers have their limitations. SIS mixers demonstrate nearly quantum limited sensitivity up to approximately 700 GHz, but then exhibit a reduction in sensitivity above 1 THz [1]. The use of HEBs are therefore utilized for operations above 1 THz, however they experience noise temperatures of approximately 10 times the quantum limit. High transition-temperature superconductor (HTS) Josephson junction (JJ) mixers have the potential for high frequency operation due to their larger superconducting energy gaps, when compared to low temperature superconductors, that are typically used in SIS receivers. HTS JJ mixers could also offer a large intermediate frequency (IF) bandwidth (10-20 GHz), low local oscillator power requirements (~100 nW), the potential for higher sensitivity at 1-2 THz, and higher temperature operation. In this work, we investigate the conversion gain of a Y-Ba-Cu-O (YBCO) focused helium ion beam Josephson junction (HI-JJ) operating as a mixer at 250 GHz.

The design of the device involves a JJ contained in a 4 µm YBCO bridge at the center of a log spiral antenna. The JJ is created with a focused helium ion beam (FHIB) which disorders oxygen in the crystal structure of YBCO, that converts the material from superconductor to insulator [2]. The FHIB technique is conducive to the THz mixer application because it allows for nano-scale junctions to be created through a trimming step with a high dose of helium ions that drives the material insulating [3]. This ultimately gives the ability to control the junction size, as well as to tune the impedance of the junction [4] to match the antenna (70 Ω).

The current-voltage characteristics (I-V) were recorded for both the unpumped and pumped case. The junction exhibited a critical current (I_c) of approximately 0.1 mA and a normal state resistance of 11 Ω. The device was pumped with 250 GHz and the I-V curve revealed several Shapiro steps corresponding to the AC Josephson effect: $V = n\frac{hv}{2e}$

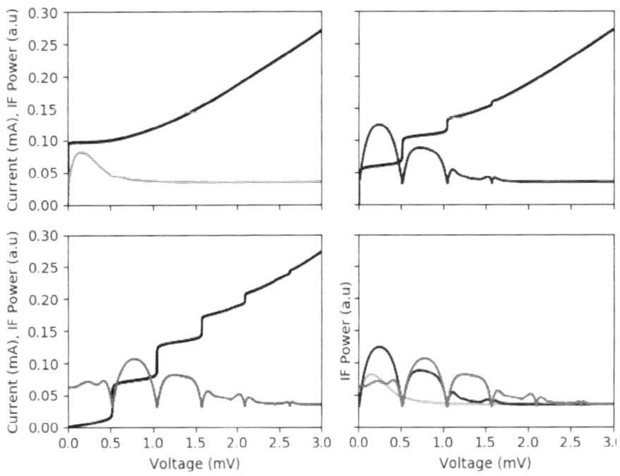

Fig. 1. I-V curves with IF power vs voltage for varying pump power at 250 GHz radiation.

where n is an integer, v is the drive frequency, and h/2e is the flux quantum. The device I-V curve and IF output power vs bias voltage was recorded with and without 250 GHz radiation for several different power levels and is shown in Fig. 1.

To determine the conversion gain of the mixer we utilize a resistively and capacitively shunted junction (RCSJ) model to find a fit to the data to approximate the incident RF power. A time dependent current source of the form $I_{rf}Sin(\omega_{rf}t)$ is added to the RCSJ model to include the AC Josephson effect. A good fit to the pumped case of half suppression of the unpumped critical current was found with I_{rf} = 0.07 mA. From this fit we determine the conversion gain of the device is approximately -5.5 dB. This is very promising when we compare it to the performance of the detectors used on the Heterodyne Instrument for the Far Infrared (HIFI), which had conversion gains ranging from -10 dB to -4 dB. This work demonstrates the great potential and flexibility of YBCO junctions directly written with a focused helium ion beam for high frequency applications.

[1]Mechanical Engineering Department, University of California Riverside, Riverside, CA, 92551 USA.
[2]Electrical Engineering Department, University of California Riverside, Riverside, CA, 92551 USA.
[3]Jet Propulsion Laboratory, California Institute of Technology, Pasadena, CA, 91109 USA.

References

[1] J. Zmuidzinas and P. L. Richards, "Superconducting detectors and mixers for millimeter and submillimeter astrophysics," *Proc. IEEE*, vol. 92, no. 10, pp. 1597–1616, 2004.

[2] S. A. Cybart *et al.*, "Nano Josephson superconducting tunnel junctions in YBa 2 Cu 3 O 7 -δ directly patterned with a focused helium ion beam," *Nat. Nanotechnol.*, vol. 10, no. 7, pp. 598–602, 2015.

[3] E. Y. Cho, Y. W. Zhou, J. Y. Cho, and S. A. Cybart, "Superconducting nano Josephson junctions patterned with a focused helium ion beam," *Appl. Phys. Lett.*, vol. 113, no. 2, pp. 29–33, 2018.

[4] A. T. Cortez, E. Y. Cho, H. Li, D. Cunnane, B. Karasik, and S. A. Cybart, "Tuning Y-Ba-Cu-O Focused Helium Ion Beam Josephson Junctions for Use as THz Mixers," *IEEE Trans. Appl. Supercond.*, vol. 29, no. 5, 2019.

Low-power consumption THz quantum-cascade VECSEL using patch-based metasurface

Christopher A. Curwen[1], John L. Reno[2], and Benjamin S. Williams[1]

Terahertz (THz) quantum-cascade (QC) lasers are attractive sources of local oscillator power for THz heterodyne receiver systems thanks to their high power and efficiency at frequencies in the 2-5 THz range [1]. In particular, THz QC-lasers have been deployed as local oscillators for the OI line at 4.74 THz. Although THz QC-lasers have traditionally been implemented in surface-plasmon or metal-metal ridge waveguides, a new architecture is the THz QC vertical-external-cavity surface-emitting laser (QC-VECSELs), which is promising due to its high-quality beam, high output power, and broadband frequency tunability [2]. The enabling component of the THz QC-VECSEL is an amplifying metasurface based on a subwavelength array of metal-metal antenna elements loaded with QC gain material. Because THz QC-lasers must be operated at cryogenic temperatures, the laser's power consumption critically determines the cryocooler load and total system power consumption. In this work, we demonstrate QC-VECSELs at ~4.7 THz based on extremely sparse metasurfaces allowing for reduced power consumption and improved temperature performance.

To date, the metasurfaces used for THz QC-VECSELs have been based on a subwavelength array of long, narrow metal-metal waveguides coupled to surface radiation via the TM_{01} cutoff resonance of the waveguides. Here, we instead use a subwavelength array of patch antennas, similarly coupled to surface radiation via their TM_{010} patch resonance. While the areal fill factor (area of QC-loaded antenna elements / total area of metasurface) of a typical ridge-waveguide based metasurface is limited to a minimum of ~15%, the fill factor of a patch array can be reduced to <5%, which reduces the required pump power for a given metasurface area, and improves temperature performance by allowing extraction of heat along two axes, instead of one.

The demonstrated device is designed to operate in a single mode at 4.63 THz and has a fill factor of 4% (47 μm periodicity along both axes of the 2-dimensional surface, and one patch per period that is 11.5 μm wide and 7.5 μm long). Additionally, the metasurface is designed with a spatially varying phase response to provide a focusing effect for improved beam quality and reduced diffraction loss [3]. The measured current, voltage, and THz power characteristics are plotted in Fig. 1. A maximum of 0.45 mW of power is measured in continuous-wave mode at 77 K, with a total power consumption of 1.25 W, for a wall plug efficiency of 0.036%.

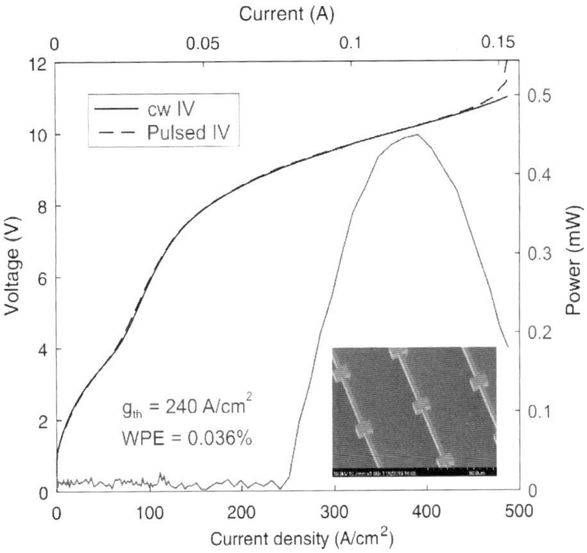

Fig. 1. Current – voltage curve of the measured devices operating at 77 K, and corresponding THz output power. Up to 0.45 mW of THz power is obtained with a wall-plug efficiency of 0.036%. The current – voltage curve measured in pulsed mode (1% duty cycle) is almost identical as that in continuous wave, indicating minimal thermal impact by operating in continuous wave mode. Inset shows SEM of fabricated metasurface.

References

[1] H. W. Hubers, *et al.*, "High-resolution terahertz spectroscopy with quantum-cascade lasers," *J. Appl. Phys.*, vol. 125, no. 15, pp. 151401, April 2019, 10.1063/1.5084105.

[2] L. Y. Xu, *et al.*, "Metasurface external cavity laser," *Appl. Phys. Lett.*, vol. 107, no. 22, pp. 221105, Dec. 2015, 10.1063/1.4936887.

[3] L. Y. Xu, *et al.*, "Focusing metasurface quantum-cascade laser with a near diffraction-limited beam," *Opt. Express.*, vol. 24, no. 21, pp. 24117-24128, Oct. 2017, 10.1364/Oe.24.024117.

[1] University of California Los Angeles, Los Angeles, CA 90095 USA.

[2] Sandia National Laboratories, Center of Integrated Nanotechnology, Albuquerque, NM, USA.

Design Study for Optimal Performance of Tunable Antenna-Coupled Intersubband Terahertz (TACIT) Mixer

Changyun Yoo[1], Jonathan Kawamura[2], Ken W. West[3], Loren N. Pfeiffer[3], Boris. S. Karasik[2], and Mark S. Sherwin[1]

Tunable Antenna-Coupled Intersubband Terahertz (TACIT) mixer is a new type of hot-electron bolometer mixer based on high-mobility 2-dimensional electron gas (2DEG) in a GaAs/AlGaAs quantum well that uses intersubband transition for efficient absorption of THz radiation [1, 2]. TACIT mixers operate at relatively high temperatures (50–70 K), and are predicted to offer low single-sideband (SSB) noise temperature (~1,000 K), a wide intermediate frequency (IF) bandwidth (~10 GHz), and low required local-oscillator (LO) power (< 1μW) along with in-situ tunability in the detection frequency (2–5 THz) [3]. These mixer characteristics potentially make TACIT mixers an attractive mixer technology for high-resolution, multi-pixel THz heterodyne instruments particularly for planetary applications.

A TACIT mixer is a four-terminal device in which two ohmic contacts (source and drain; see Fig. 1. (a)) are used to read out the IF response due to the change of the device resistance caused by THz radiation, and top and bottom gates are used to couple THz radiation into the active region of the device (see Fig. 1. (b)). A planar THz antenna integrated in the top and bottom gate metallization orients THz fields perpendicular to the 2DEG plane at the active region, in which THz energy is resonantly absorbed by the 2DEG through an intersubband transition. The excited electrons quickly thermalize and heat up the active region, resulting in a fast bolometric response in the device resistance. The top and the bottom gates can also electrically tune the intersubband absorption frequency (via the dc Stark effect) as well as the device impedance seen by the THz antenna.

In our recent work [2], we have successfully fabricated a prototype TACIT mixer for 2.5 THz using a flip-chip technique to perform lithography on both sides of a sub-micron thick quantum well membrane, and have observed tunability in the detection frequency (2.5–3.1 THz) and THz mixing at 60 K with an IF bandwidth exceeding 6 GHz. However, the antenna structure and the 2DEG mesa geometry of the prototype device were not designed for optimal mixer performance, and thus the noise temperature, the conversion gain, and the required LO power could not be characterized in the current prototype device.

[1] Physics Dept. & Inst. for Terahertz Science and Technology, University of California at Santa Barbara, Santa Barbara, CA 93106, USA.

[2] Jet Propulsion Laboratory, California Institute of Technology, Pasadena, CA 91109, USA.

[3] Dept. of Elec. Eng., Princeton University, Princeton, NJ 08544, USA.

To achieve optimal mixer performance of a TACIT mixer, a new design and fabrication effort is under way. In this work, we theoretically or experimentally investigate all design parameters of a TACIT mixer including the 2DEG mesa geometry and antenna design and evaluate how they affect the mixer performance. Details on the design and estimated mixer characteristics will be provided in the presentation.

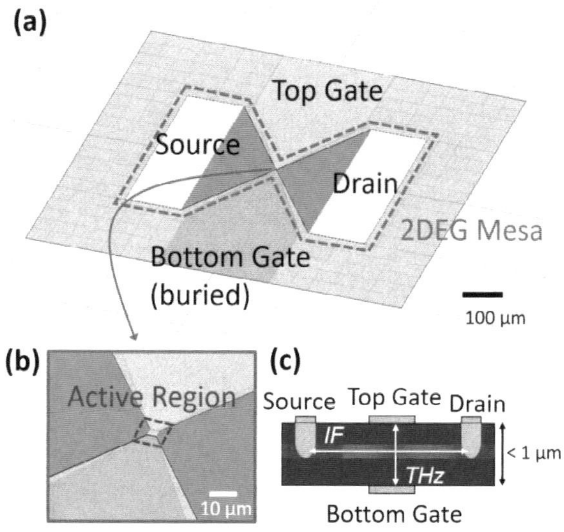

Fig. 1. TACIT mixer integrated with a bow-tie antenna for optimal mixer performance (a) layout for a TACIT mixer showing 2DEG mesa (outlined by green dashed lines) and a bow-tie antenna integrated in top and bottom gates, (b) active region of a TACIT mixer, (c) vertical schematic of a TACIT mixer.

References

[1] M. S. Sherwin, "Tunable Antenna-Coupled Intersubband Terahertz (TACIT) Detector," US Patent 5,914,497, Jun. 22, 1999.

[2] C. Yoo et al, "Demonstration of a Tunable Antenna-Coupled Intersubband Terahertz (TACIT) Mixer." *Appl. Phys. Lett* (in-press)

[3] M. S. Sherwin et al, "Tunable Antenna Coupled Intersubband Terahertz (TACIT) Mixers: The Quantum Limit Without the Quantum Liquid," *Proc. Far-IR, Sub-mm and mm Detector Technology Workshop*, 2002.

NOEMA heterodyne receivers performance and tuning optimization

Christophe Risacher[1] with the IRAM Frontend group

The NOEMA interferometer located in the French Alps at 2500m altitude, currently operates 10 antennas having 15m of diameter with the 11th antenna coming out in April 2020 and the 12th and last one in 2021.

The receiver cabin is equipped with a single closed cycle cryostat (see Fig. 1), containing 4 heterodyne frequency bands, covering the 70-375 GHz frequency range. It is cooled down using a 3-stage GM cooler from Sumitomo Heavy Industries, Ltd. The end temperature is typically below 3K, ensuring the SIS junctions operate in the 3.5-4.5K temperature range.

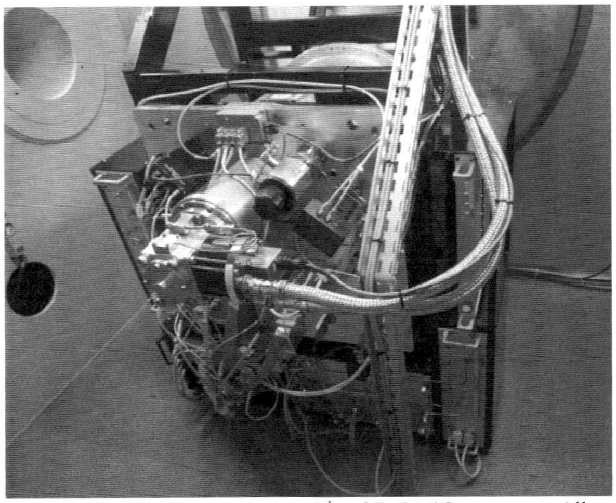

Fig. 1. Receiver installed in the 10th NOEMA 15m antenna. All critical electronics is directly mounted on the top plate.

We will present an overview and summary of the receivers performance gathering results after several years of operations. We will also show the statistics for the receiver tunings over the past years assessing tuning quality, tunings which required manual intervention and finally total required tuning time.

As observing time is very expensive, any gain in improving on source observing efficiency is beneficial. Therefore, we will also show improvements and optimizations of the automated tuning procedures performed in 2019 allowing to further improve the overall tuning reliability, quality and total duration. Finally, we will present the next planned steps to further improve it.

[1] IRAM, 300 rue de la Piscine, 38406 Saint-Martin-d'Heres, France

Development and Testing of the 1.46 THz and 1.9 THz GUSTO Flight-Model Local Oscillator Arrays

Jeffrey L. Hesler[1], Steven Retzloff[1], Corey Gardner[1], Silvio Mancone[1], Ben Swartz[1], Cliff Rowland[1], and Thomas W. Crowe[1]

This talk will describe recent work on the 1.46 THz and 1.9 THz local oscillator (LO) arrays for the Galactic/Extragalactic ULDB Spectroscopic Terahertz Observatory (GUSTO) project [1]. This project is led by the University of Arizona, and consists of a long duration balloon astronomy mission to be launched from Antarctica. VDI is developing 2x4 pixel arrays of LOs at 1.46 THz and 1.9 THz that will be used to drive 2x4 pixel HEB arrays. Each LO pixel consists of a stand-alone amplified multiplier chain, that had been previously developed. These chains have been re-packaged to fit the requirement of a 15 mm spacing between pixels. A photograph of a 1x4 array of the 1.9 THz local oscillator pixels is shown in Fig. 1.

Full 2x4 arrays of both frequencies have been tested extensively to verify that they meet the flight requirements. First, a minimum of 15 uW output power is required for each pixel over the frequency range. Also, the THz power must be electronically controllable over a 6 dB range of power. In addition, the amplitude noise of each pixel is measured to insure that the amplitude noise stays below the specified -147 dBc/Hz at both full power and at minimum power. The inset in Fig. 1 shows a typical curve for a 1.9 THz pixel at full output power, with a maximum noise of -157 dBm. This measurement uses a method previous developed for the ALMA local oscillators [2]. Finally, the chains are tested over an operational temperature range of 0C to close to 50C, to verify proper performance over this range.

The results of these test for the two arrays will be presented at the symposium.

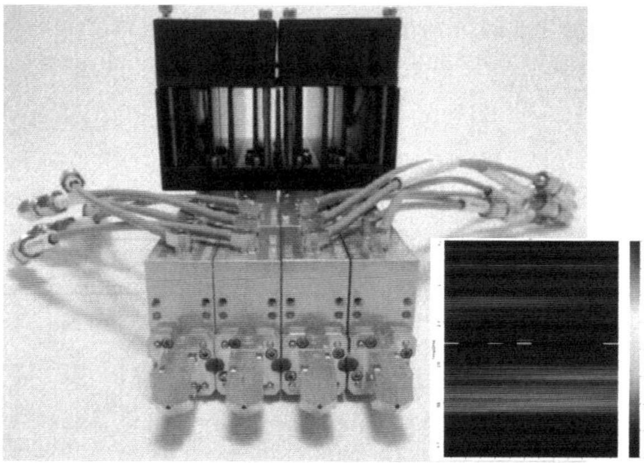

Fig. 1. Photograph of a 1x4 array of 1.9 THz local oscillators. The inset shows an amplitude noise measurement of one of the pixels over the required frequency band. The dark blue indicates -170 dBm/Hz and the light blue peaks are at -157 dBm, well below the specified -147 dBm.

REFERENCES

[1] https://www.as.arizona.edu/gussto

[2] E. Bryerton and J. Hesler "Sideband noise screening of multiplier-based Sub-millimeter LO chains using a WR-10 Schottky mixer", Proc. Of 19th Int. Symp. on Space Terahertz Tech., 2008, pp. 498-501.

[1] Virginia Diodes Inc., Charlottesville, VA, 22902 USA .

Compact Multi-Pixel Frequency Multiplied Local Oscillator Sources for Wideband Array Receivers in the 200-600 GHz & 1.4-2.7 THz Ranges

Jose V. Siles, Jonathan H. Kawamura, Robert H. Lin, Choonsup Lee, Alain E. Maestrini
Darren J. Hayton, Ken B. Cooper, Maria Alonso del Pino and Imran Mehdi.

Following on the success of high-spectral resolution submillimeter-wave/terahertz receivers recently flown in space, such as MIRO on-board the Rosetta Orbiter and HIFI on-board the Herschel Space Observatory, the next generation of terahertz cameras needs to address the challenge of achieving broadband multi-pixel operation within the strict power budgets of space missions. For planetary science applications, the 200-600 GHz range is key for studying water (and its composition) in comets and ocean worlds, in the search for habitability and the origin of water on Earth. Fast mapping is key in order to be able to avoid the data retrieval issues caused by the fast rotation of comets and to fulfill the science goals during flyby missions to the outer planets and their moons. For Astrophysics, large-band array receivers in the 1.4-2.7 THz are required to capture all key tracers of star formation: [CII], [NII], [OI], CO, HD, etc. Last but not least, dc power consumption is one the main limiting factor for terahertz array receivers. Since the local oscillator (LO) subsystem mostly drives the total power consumption of such receivers, ultra-compact high-efficiency LO systems are required to maximize the number of pixels affordable for future NASA missions involving THz spectroscopy.

Herein, we report in the latest progress towards low dc-power compact multi-pixel Schottky diode-based frequency multiplied local oscillator sources operating at room temperature. For the 200-600 GHz range, progress towards an ultra-broadband 16-pixel 210-600 GHz local oscillator array with >2 mW output power per pixel and ~20 W total power consumption (~1,25 W/pixel), enough to drive Schottky diode-based mixers with healthy margins. For the 1.4-2.7 THz array sources, we will report on the development status of a new 16 pixel 1.9-2.06 THz source with >20 μW/pixel and ~26 W total power consumption (~1.6 W/pixel). This represents one order of magnitude improvement with regards to the JPL 1.9 THz single-pixel LO flown on HIFI (~25 W/pixel) and the state-of-the-art JPL 4-pixel LO flown on STO-2 (20 W/pixel). We will also report on the 1st multi-pixel local oscillator in the 2.5 THz region (see Fig. 1.). This 2.5 THz source produces 15-37 μW/pixel with than 10 W dc power consumption. Currently, the most powerful 2.5 THz LO source flown on GREAT/SOFIA provides ~3-5 μW at 300K. The newest JPL single-pixel source at 2.5 THz produces up to 50 μW and is approximately one-fourth in size than the previous generation. Likewise, the newest 1.6 THz source developed at JPL provides ~0.7 mW (x10 than the previous JPL source in this range). Using phase grating techniques, such a chain could drive at least 64-pixel HEB-based receivers.

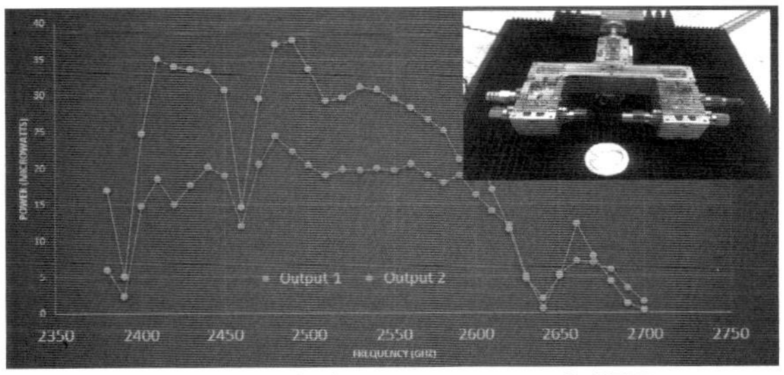

Fig. 1. First ever 2-pixel room-temperature 2.5 THz source.

These improvements in output power and reduction in dc power are mostly due to the JPL patented on-chip power combined topology and thoroughly optimized frequency multipliers [1]. An overview of the latest results will be provided, showing that the efficiency of this new generation of frequency multiplied sources follows now very closely the predictions by accurate physics-based models.

Achieving larger RF bandwidths (up to 50%) for terahertz room-temperature sources is one of the 'unexplored' territories where we are concentrating our efforts on now. For future balloon-borne instruments (e.g. ASTHROS, planned to fly from Antarctica in 2023) and space-borne (e.g. HERO, one of the instruments under study for the Origins Space Telescope concept), this kind of frequency coverage is necessary to maximize science return while minimizing the number of receiver channels to reduce complexity, dc power and cost. Progress towards enhancing the RF bandwidth of these LO sources by a factor of two in the 210-600 GHz and the 1.4-2.7 THz bands will be discussed.

References

[1] J. V. Siles, et al., "A New Generation of Room-Temperature Frequency Multiplied Sources with up to 10x Higher Output Power in the 160 GHz-1.6 THz Range", IEEE Transactions on Terahertz Science and Technology, Nov. 2018.

[1] NASA Jet Propulsion Laboratory, California Institute of Technology, Pasadena, CA 91109, USA.

The research was carried out at the Jet Propulsion Laboratory, California Institute of Technology, under a contract with the National Aeronautics and Space Administration.

Stabilization of terahertz quantum-cascade lasers by near-infrared optical excitation

M. Wienold[1], T. Alam[1], X. Lü[2], K. Biermann[2], L. Schrottke[2], H. T. Grahn[2], and H. W. Hübers[1, 3]

Frequency and power stability is an important requirement for a local oscillator (LO) in a heterodyne spectrometer. For terahertz (THz) quantum-cascade lasers (QCLs) employed as an LO, passive stabilization can be realized by keeping the driving current and operating temperature as stable as possible [1]. In this way, an LO frequency stability of a few MHz in a mechanical cooler is achieved for the 4.75-THz channel of the German Receiver for Astronomy at THz frequencies (GREAT). This is sufficient for typical astronomic observations. However, for resolving very narrow spectral details in the atmospheres of the Earth and other planets, a linewidth of less than 1 MHz is desirable. Besides frequency stability, power stability of the QCL is important, since the pump power stability mainly limits the Allan time for bolometric mixers.

Previously, we have demonstrated how near-infrared excitation can be exploited for light-induced frequency tuning of THz QCLs [2]. Here, we show how this effect in combination with the tuning of the QCL driving current can be used to realize simultaneous frequency and power stabilization. For the experiments, a QCL emitting at 3.1 THz is mounted in a mechanical cryocooler (Ricor, K535). An optical fiber is positioned close to the QCL for near-infrared illumination. A methanol gas cell, a Ge:Ga photoconductive detector, and a lock-in amplifier with integrated PID capabilities (ZI, UHF) are used for modulation (1f) spectroscopy and frequency locking. A second Ge:Ga photoconductive detector is used for power monitoring and, in combination with a second PID loop, for power locking. Figure 1 depicts experimental results for such an active stabilization, where the QCL driving current is used for frequency locking and the near-infrared illumination for power stabilization. In the free-running case (1), the frequency varies by almost 20 MHz associated with cooler vibrations, and the output power varies by 0.1% [root-mean square value (rms)]. For simple frequency locking (2), frequency variations are reduced to 240 kHz [full width at half maximum (FWHM)], while output power variations increase to 0.4% (rms). In case of simultaneous frequency and power stabilization, variations of 260 kHz (FWHM) and 0.03% (rms) are obtained, respectively. More details can be found in [3]. The advantage of our method is that both parameters (frequency and output power) can be controlled on the same short time scale, since the related current and illumination tuning effects are physically fast processes. This allows for an efficient compensation of detrimental power variations in case of frequency locking and vice versa. The approach might be adopted for a heterodyne receiver by using a subharmonic mixer and a multiplied microwave reference for frequency locking.

Fig. 1: (a) Frequency and (b) output power stability for the three cases: free running, frequency locked, frequency and power locked.

[1] German Aerospace Center (DLR), Institute of Optical Sensor Systems, Rutherfordstr. 2, 12489 Berlin, Germany

[2] Paul-Drude-Institut für Festkörperelektronik, Leibniz-Institut im Forschungsverbund Berlin e. V., Hausvogteiplatz 5–7, 10117 Berlin, Germany.

[3] Humboldt-Universität zu Berlin, Department of Physics, Newtonstr. 15, 12489 Berlin, Germany.

References

[1] H. Richter, M. Wienold, L. Schrottke, K. Biermann, H. T. Grahn, and H.-W. Hübers, "4.7-THz local oscillator for the GREAT heterodyne spectrometer on SOFIA," *IEEE Trans. Terahertz Sci. Technol.* vol. 5, pp. 539–545, Jul. 2015.

[2] T. Alam, M. Wienold, X. Lü, K. Biermann, L. Schrottke, H. T. Grahn, and H. W. Hübers, "Wideband high-resolution terahertz spectroscopy by light-induced frequency tuning of quantum cascade lasers," *Opt. Express* vol. 27, pp. 5420–5432, Feb. 2019.

[3] T. Alam, M. Wienold, X. Lü, K. Biermann, L. Schrottke, H. T. Grahn, and H. W. Hübers, "Frequency and power stabilization of a terahertz quantum-cascade laser using near-infrared optical excitation," *Opt. Express* vol. 27, pp. 36846–36854, Dec. 2019.

NOTES:

Advances in High-Power THz Sources

Steven Retzloff[1], Eric Bryerton[1], Jeffrey Hesler[1], and Thomas W. Crowe[1].

Solid-state sources based on amplifier-multiplier-chains (AMCs) continue to serve as the source of choice for radio astronomy and atmospheric sensing. As the available output power continues to increase, other applications are also being enabled. These include plasma diagnostics, ESR, DNP-NMR, and radar imaging systems. This paper will review recent developments at VDI that are leading to record power levels as well as more reliable and compact sources.

As an example, Fig. 1 indicates the output power from a ~260 GHz source; roughly 250 mW across a useful band of 10 GHz. This result is made possible through the use of diode multipliers optimized for high power handling, in terms of both electrical and thermal design. Two-way in-phase, power combining is also used for the final two frequency doublers. Significant further improvements are anticipated with higher levels of power combining and/or the cooling of the components (for applications where cooling is feasible). Sources such as these are also useful to drive multiplier into the range well above 1 THz, and data for these higher bands will be presented at the conference.

Figure 2 shows the power from a similar source using broadband components. In this case, roughly 100mW is available across the entire WR10 waveguide band. This performance has been enabled through the use of a newly available amplifier technology. The power available from these broadband sources can also be multiplied to higher frequency, resulting in the results shown in Fig. 3.

The potential for further improvements in this source technology will be summarized at the conference.

Fig. 1. The measured power of a VDI source. The multiplier bias and amplifier power are optimized at each point. (All power levels measured with PM5 power meter.)

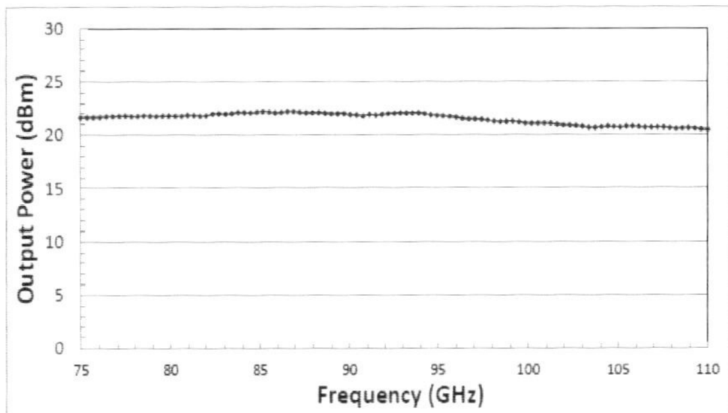

Fig. 2: Power from a broadband (WR10.0) source.

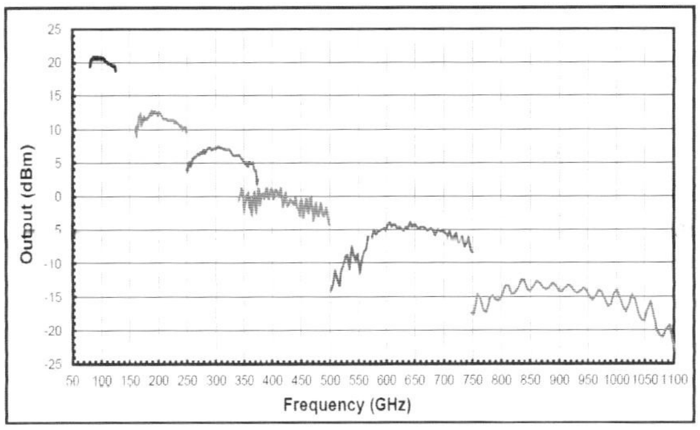

Fig. 3: Power from a broadband modular (reconfigurable) source using a driver module (similar to Fig. 2) and a standard set of frequency multipliers.

[1] Virginia Diodes Inc., Charlottesville, VA 22902 USA.

NOTES:

Millimeter-Wave Filter Bank Spectrometers

Kyle Massingill, Sean Bryan, Christopher Groppi, Philip Mauskopf, Bianca Pina, Philip Rybak, and Peter Wullen

Abstract—**We present testing of a 180GHz prototype filter bank spectrometer and report on designs for larger channeled millimeter-wave filter banks at 50-183 GHz. This technology will be utilized in the development of a novel low-SWaP-C microwave sounding sensor. It works by amplifying the broadband signal with an LNA then channelizing with the millimeter-wave filter bank. Each channel of the filter bank is detected by a separate diode. The sensor system enabled by the millimeter-wave filter bank has a great potential for measuring 3D atmospheric water vapor and temperature by detecting the 183 GHz water line. We will report here our progress in utilizing the sensor for on the ground detections of humidity and preliminary designs for the use of the sensor on a small satellite. Satellite based microwave radiometers are the most important driver of global weather forecasting. Current sensors rely on high-Swap-C LO/mixer/RF components. We will demonstrate that the maturation and adoptions of our novel sensor system would greatly improve weather forecasting.**

The prototype millimeter-wave filter banks were fabricated using a micro milling CNC with a tolerance of 5-microns and is a waveguide coupled to five spectrometer channels. The spectrometer channels are resonant with a tuned center frequency. We tested a single channel prototype to determine sensitivity. Testing was performed by driving a VNA extender with a signal generator to input a sweeping signal into the prototype.

I. Introduction

DATA from microwave radiometers on large U.S. weather satellites is the single highest impact driver of global weather forecasting [1] [2]. These satellites carry instruments such the Advanced Technology Microwave Sounder (ATMS) and Advanced Microwave Sounding Unit (AMSU). These sensors use heterodyne mixer technology, followed by an RF filter bank and detectors. This effective approach comes with a relatively high cost in SWaP-C and system complexity. First, it requires a millimeter-wave local oscillator, which uses significant power and requires an additional temperature (and possibly frequency) stabilization system. After down conversion by a mixer, in existing systems the channels are separated with an RF filter bank which inherently limits instantaneous bandwidth and may be less compact due to longer RF wavelengths. Finally, existing mixers do not perform sideband-separation without additional complexity, meaning that signals from frequencies both above and below the line are combined. This complicates data analysis and increases the calibration and local oscillator stabilization requirements.

These disadvantages of existing spectrometer systems drove us to develop millimeter-wave waveguide filter bank technology [3] at Arizona State University, forming the heart of our novel approach. Our approach, as illustrated in the conceptual overview in Figure 1, is to first amplify the signal from the scene with a commercial millimeter-wave low noise amplifier (LNA). The signal then goes to a millimeter-wave filter bank that we fabricate at the ASU precision machining facility. The filter bank consists of resonant cavities that couple each frequency channel to a different output waveguide port. A millimeter-wave diode power detector at each output port detects the signal which a commercial instrumentation amplifier IC then amplifies before digitization, and storage. Our key enabling technology, the waveguide filter bank, has been successfully tested at 90 GHz [3] and 180 GHz [4] in the laboratory and in some integrated tests.

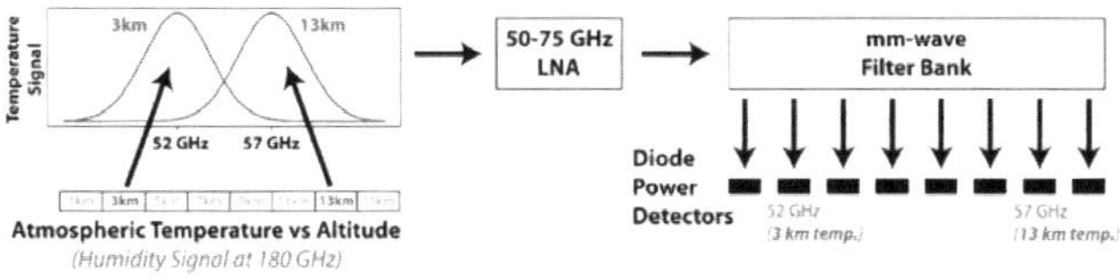

Fig. 1. Conceptual overview the spectrometer. Temperature (bottom left) and humidity signals from different atmospheric altitudes appear at different millimeter wave frequencies (top left) and are measured separately by diode detectors (bottom right) at each output of the filter bank.

A. Pressure Broadening

The 183GHz water molecule spectral line is commonly used in weather satellites to observe vertical profiles of water vapor in the atmosphere. Pressure broadening causes the emission of water vapor in lower (and higher pressure) parts of the atmosphere, to be "broadened" several GHz away from the 183GHz rest frequency. While emission from water vapor in the higher (and lower pressure) parts of the atmosphere stays close to 183GHz. The broadening is caused as a resulted of molecular collisions that are more likely at higher pressure and temperature. As illustrated in figure 3, observations several GHz away from the 183GHz line has emission only from the lower part of the atmosphere while observation from near 183GHz has emission from both the higher and lower parts of the atmosphere. Retrieving the humidity profile from the measured spectrometer data is a non-linear fitting problem, or inverse problem [5]. Pressure broadening, analyzed with careful radiative transport simulations of the atmosphere, allows the radiometer measurements to be converted to measurements of atmospheric temperature. This is demonstrated in existing codes such as the am code [6].

II. Design

A. Spectrometer

To meet the goal of delivering a low volume, low power, and high channel count millimeter-wave atmospheric spectrometer,

Submitted for review 1 March 2021

K. Massingill, C. Groppi, P. Mauskopf, B. Pina, P. Rybak, and P. Wullen are with the School of Earth and Space Exploration, Arizona State University, Tempe, AZ 85287 USA (e-mail:kyle.massingill@asu.edu)

S. Bryan is with School of Electrical, Computer, and Energy Engineering, Arizona State University, Tempe, AZ USA

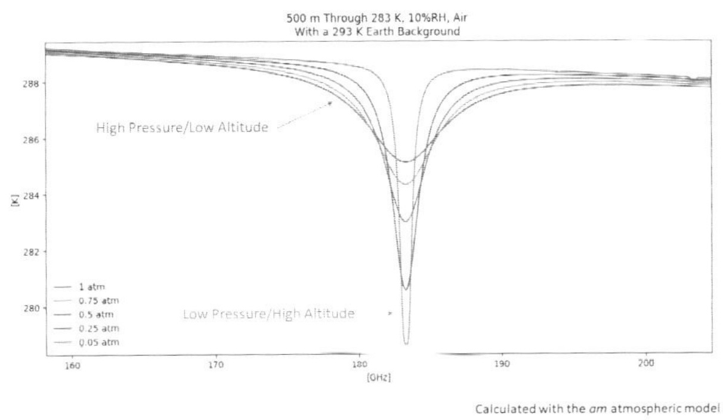

Fig. 2. Demonstration of pressure broadened signal from water vapor 183GHz spectral line as determined by atmospheric simulation. [6]

we have designed the following system. As illustrated in figure 3, light enters through a pyramidal feed horn, then is amplified at broadband by two LNAs. Frequencies are then selected off as the light travels through the filter-bank. Each channel of the filter-bank terminates on a diode detector. The LNAs are the only part of the spectrometer that require power, and the instrument can be made exceedingly compact.

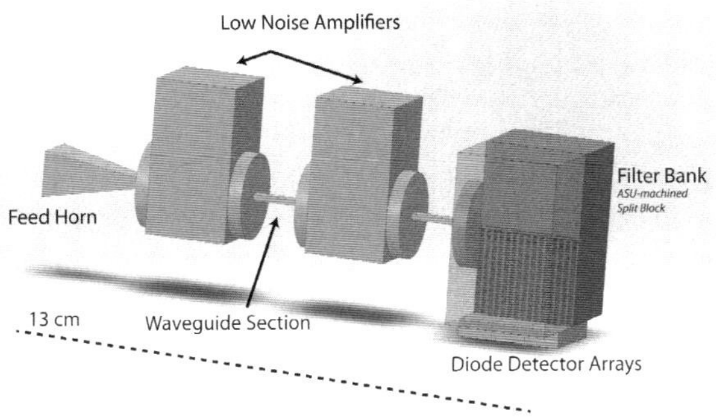

Fig. 3. Drawing of spectrometer system. Broadband signal enters the feed horn, is then amplified by LNAs and finally divided by filter-bank. The filter-bank is a passive component of directly milled metal based on a novel design.

B. Filter-Bank

The primary novel technology of our mm-wave spectrometer is the filter-bank. The design for mm-wave filter-banks presented here was developed at the ASU Astronomical Instrumentation Lab and are described in more detail in [3] and [4]. The filter-bank consists of a primary rectangular waveguide and tee coupling channels tuned to specific frequencies. The channels are tuned by narrow coupling section then a half-wavelength resonating cavity. A second narrow section on the other end of the resonant cavity, defines the length of resonance. The channel is then terminated by a detector. The narrow sections have a cutoff frequency 50% higher than the center frequency of the channel. The center frequency of each channel is defined by the length of the resonant cavity, while the bandwidth is defined by length of the narrow cutoff sections. An example filter-bank is shown in figure 4. Multiple five channel examples of this filter-bank have been designed and modeled in HFSS with full 3D simulation, that include effects from machining such as rounded corners on channels. Larger channeled filter-banks have been modeled using a cascading S-matrix approach. This is achieved by 3D simulating each channels individually to get the 3 port S-matrix then cascading the S-matrix of all the channels. This has been done for a 54 channel filter with pass bands 135-170GHz and 190-245GHz. Two prototypes of five channel filter-bank designs have been fabricated and tested. The prototypes were drilled in aluminum on a 5-micron tolerance CNC mill. One was made in WR10 with for 80-105GHz [3], the other was made in WR5 for 156-203GHZ [4]. Testing was completed by sweeping over injected frequency using a signal generator and recording the voltage induced on diodes terminating the channels. The results show broad agreement between simulation and testing.

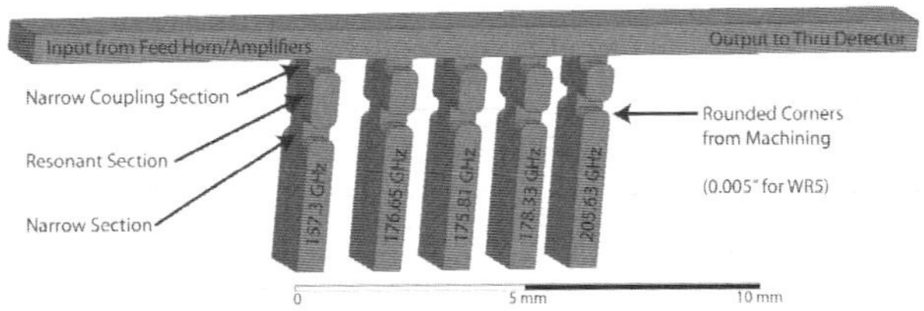

Fig. 4. An example of the waveguide filter-bank concept. Broadband light propagates through the waveguide and different frequency bands are selected off of the main waveguide by the five channels.

III. Testing

We have demonstrated our spectrometer technology with a single-channel prototype system in a laboratory environment. For this demonstration, we connected two commercial Radiometer Physics LNAs to the input port of the prototype spectrometer tested in [4], and connected a Pacific Millimeter Products GD diode detector to the 156 GHz output port. We used laboratory SRS amplifiers to measure the detector signal. We viewed ambient temperature and 77 K liquid nitrogen loads to calibrate the system, then took over a minute of data viewing the ambient temperature load to measure the noise. We clearly observed the temperature signal in the radiometer, and the measured noise performance shown in Figure 5. This shows that the system is stable with no excess (i.e. 1/f) noise detected down to the lowest clearly resolvable audio frequency ($\sim$ 50 mHz, or $\sim$ 20 seconds). This demonstrates that our novel spectrometer maintains calibration and stability over the long timescales required in microwave sounding observation.

Motivated by this successful laboratory demonstration, and to set requirements for the amplifiers, detectors, and other components, we developed a noise budget for the full sensor system. First considering only the channel bandwidth and the system temperature, the per-channel sensitivity $NE(\triangle)T$ is set by the radiometer equation

$$NE(\triangle)T = (T_{scene} + T_{sys})/\sqrt{BW}, \tag{1}$$

where T_{scene} is the incident signal and BW is the channel bandwidth. Including noise contributions from other system elements the noise is calculated to be $30.5\ \mathrm{mK}\sqrt{\mathrm{s}}$, close to the fundamental limit. For the 60 GHz channels, the forecast is $40.5\ \mathrm{mK}\sqrt{\mathrm{s}}$. Our measured noise level of $230\ \mathrm{mK}\sqrt{\mathrm{s}}$ is close to the $80\ \mathrm{mK}/\sqrt{\mathrm{s}}$ level our model forecast for our prototype, with the difference due to insufficient gain in the prototype amplifiers. This will be corrected easily in the fielded instrument. Our laboratory demonstration proves the feasibility of a radiometer based on waveguide filter bank technology.

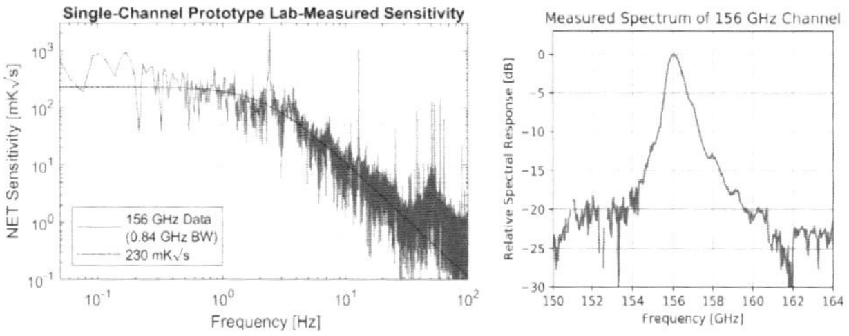

Fig. 5. Measured sensitivity (left) and passband (right) of a single-channel prototype. This demonstrates the feasibility of our approach, and has excellent noise stability (< 50 mHz, > 20 s).

IV. Conclusion

The CubeSounder mission was recently selected by the NASA flight opportunities program to mature and flight test this technology. We will fly a prototype spectrometer system on a suborbital balloon platform to demonstrate the ability of the technology to detect water vapor in the relevant environment.

At the start of the program, we will use our atmospheric simulation pipeline to finalize the selection of frequency channels and bandwidths to achieve optimum temperature and humidity sensitivity. This will in turn let us finalize the design of the detector readout electronics, and finalize the design of the 60 GHz filter bank. This will lead to final circuit designs and mechanical drawings of the payload, concluding with an internal design review before fabrication. We will build the flight control, power, and detector readout electronics. We will also write flight control software for the embedded control PC, and integrate the payload. Our first flight will use an existing 183 GHz spectrometer and LNAs to enable our first flight to come relatively early in the program. The second flight will re-use the recovered flight and detector readout systems, and we will then integrate the 60 GHz sensor.

System testing in the laboratory will consist of using an ambient temperature/liquid nitrogen calibration target to measure radiometer sensitivity and responsivity. At ASU, we already have several Vector Network Analyzers (VNAs) and VNA extensions up to 210+ GHz which we will use to measure the optical efficiency of the spectrometer and the spectral passband of each channel. Similar test results from our single-channel prototype are shown in Figure 5. After verifying the sensitivity of the integrated sensor payload, we will conclude the testing by observing the atmosphere above the ASU campus. To verify that our system will operate in the balloon environment, ASU has a large thermal vacuum testing facility that has been successfully used to test space and near-space hardware.

Flight testing is the key next step for CubeSounder technology. While laboratory testing has already shown the high performance of our sensor technology, ground testing has significant limitations. Crucially, ground testing does not enable us to conclusively show that our technology performs well in the balloon flight environment, and only permits limited testing of system integration and autonomous operation. Also, ground testing only permits imaging of a small part of the atmosphere, limiting our ability to verify our sensor data by comparing with known weather conditions.

References

[1] C. Cardinali, "Monitoring the observation impact on the short-range forecast," *Quarterly Journal of the Royal Meteorological Society*, vol. 135, no. 638, pp. 239–250, 2009.

[2] E. National Academies of Sciences and Medicine., *Thriving on Our Changing Planet: A Decadal Strategy for Earth Observation from Space*. Washington, DC: The National Academies Press, 2018. [Online]. Available: https://www.nap.edu/catalog/24938/thriving-on-our-changing-planet-a-decadal-strategy-for-earth

[3] S. Bryan, G. Che, C. Groppi, P. Mauskopf, and M. Underhill, "A Compact Filter-Bank Waveguide Spectrometer for Millimeter Wavelengths," *IEEE Transactions on Terahertz Science and Technology*, vol. 5, no. 4, pp. 598–604, 2015.

[4] S. Bryan, J. Aguirre, G. Che, S. Doyle, D. Flanigan, C. Groppi, B. Johnson, G. Jones, P. Mauskopf, H. McCarrick, A. Monfardini, and T. Mroczkowski, "WSPEC: A Waveguide Filter-Bank Focal Plane Array Spectrometer for Millimeter Wave Astronomy and Cosmology," *Journal of Low Temperature Physics*, vol. 184, no. 1-2, pp. 114–122, 2016.

[5] A. Deepak, "Inversion Methods in Atmospheric Remote Sounding - NASA-CP-004," Tech. Rep., 1977. [Online]. Available: http://hdl.handle.net/2060/19780004643

[6] S. Paine, "The am atmospheric model," 2018.

Development of a Tone-Tracking Algorithm for Maximizing Dynamic Range of Kinetic Inductance Detectors

Jonathan R. Hoh[1], Adrian K. Sinclair[1], and Ryan C. Stephenson[1]

Abstract— **Here we present the designs and preliminary results of an FPGA-based tone-tracking firmware that always maintains on-resonance probe tones in frequency multiplexed MKID arrays. The design of the firmware relies on the fact that resonators have a distinct phase shift at resonant frequency that remains invariant even as the resonance itself changes.**

Similar to proportional/derivative/integration (PID) digital controllers, the firmware compares phase feedback loops to resonant phase data collected before beginning observations. Using these phase comparisons, the tone-tracking algorithm can individually update each tone in the comb sent to an MKID array so that resonance is always known and maintained. In practice, this eliminates the need for mid-observation VNA sweeps and allows MKID-based cameras to reach their maximum dynamic range potential.

The application of this tone-tracking firmware will drastically change the approach for reading out MKID arrays. We will conclude with a discussion on how other aspects of the readout system can be modified and simplified to accommodate tone-tracking along with future NASA missions that hope to use the design. Finally, we look forward at the possibility of implementing tone-tracking firmware on the cutting-edge Xilinx RFSoC FPGA.

I. Introduction

Over the new millennium, microwave kinetic inductance detectors (MKIDs) have become a popular technology for use in mm and sub-mm astronomy. MKIDs function by cryogenically cooling certain metals at superconducting temperatures in order to create a resonator which changes in both resonant frequency and quality factor when exposed to incoming photons. By creating an array of these MKIDs, each with a unique resonant frequency, the entire system can be read out with a single transmission line using a method known as frequency multiplexing. Depending on the arrangement MKIDs can be used as pixels for both spatial and spectral deep-space cameras.

In order to frequency multiplex an MKID array, a comb of tones comprised of resonant frequencies must be sent down a single transmission line coupled to all the detectors. By reading the transmitted spectral phase and amplitude information, incident power on each of the MKID pixels can be determined. The simplicity of frequency multiplexing an MKID array is one of the major upsides of using the technology versus competing detectors such as Transition Edge Sensors (TES) bolometers.

Recently, the cutting-edge technology of MKIDs and the associated frequency multiplexing techniques have been flown on balloon missions such as OLIMPO and BLAST-TNG with reasonable success. The BLAST-TNG mission in particular managed to capture images of a far-infrared stellar source using MKID detectors before the early demise of the flight. The successful implementations of MKID technology as a means of ultra-high sensitivity detection shows great promise for the future of mm and sub-mm astronomy. Unfortunately, the growing use of MKIDs for astronomy has also illuminated certain problems with their implementation. Current readout techniques for MKIDs greatly reduce their theoretical dynamic range; an issue that needs to be remedied if KIDs are to continue their widespread use.

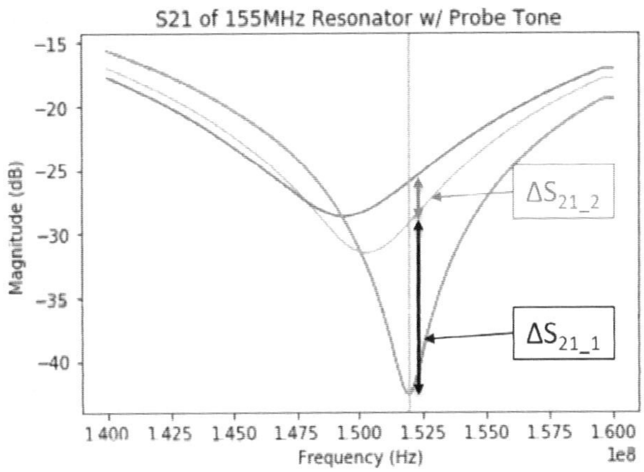

Fig. 1. An exaggerated example depicting the diminishing returns of incident power on a KID. Increasing power returns smaller and smaller changes in amplitude until the difference between two modulated states is dominated by amplifier noise. As of now, the only correction possible is a time-intensive mid-flight VNA sweep.

The current readout technique for MKID arrays as stated earlier is through frequency-domain multiplexing (FDM). By tuning each element of the array to a slightly different resonant frequency, the phase and magnitude of each detector can be individually read out from the same signal line. For small amounts of incident power, this method of monitoring has

[1] The Arizona State University, Tempe, AZ 85282 USA (contact info: jrhoh@asu.edu)

shown to be perfectly acceptable. However, problems begin to arise when there are very large changes to the incident power on the detectors.

As the intensity of light interacting with a KID increases, the change in amplitude seen at port 2 (the output to the readout FPGA) begins to shrink rapidly. Once a KID is saturated with enough power, the change in transmission is small enough that it is lost in system noise, mainly due to amplifiers. Fig. 1 shows an exaggerated example of this difference in visibility. Attempting to read out useful information from a KID which is dominated by amplifier noise is a fruitless effort and the detector is dead weight until the issue is corrected. Currently, the only remedy for this problem is a time-consuming VNA sweep which occurs mid-flight and wastes precious observation time. Losing amplitude information to amplifier noise is not uncommon during flights and on the recent BLAST-TNG mission this problem occurred every few degrees when scanning the sky.

The next NASA mission planning to use MKIDs as a means of deep-space detection is the Terahertz Intensity Mapper (TIM) balloon-borne telescope, led by a group at the University of Illinois in collaboration with a number of other research institutions. TIM aims to survey CII, OI, and OIII in 100 galaxies to demystify many aspects of star formation while simultaneously producing deep tomographic maps of the 3-D structure of the universe. The architecture, spectrometer, and readout system of TIM is modeled from the BLAST-TNG mission that recently flew at McMurdo Base in 2019. Among the lessons learned from BLAST-TNG was the need for a more efficient method to overcome the KID dynamic range issue.

II. MATERIALS AND METHODS

The solution to the restricted dynamic range lies in the development of an FPGA-based firmware that constantly keeps track of the resonance frequencies of the MKIDs within an array. Therefore, rather than producing a static probe tone for each detector that is reset once information is lost, the tone-tracking firmware will constantly update the frequency of the probe tone to assure that it is always on resonance. This tone-tracking firmware can be accomplished with a feedback control system.

The most important aspects of most control systems, known as the "invariant," is a variable which in this case is independent of the incident power on the detectors. Invariant properties allow for a reference during feedback loops which allow the system to change relative to a stable value. Resonators have such a property due to their electrical length at resonance being zero. This means that regardless of incident power on a KID, the phase of a tone at resonance will never change (ignoring linear phase shifts from signal lines connected to the system).

Since the goal of resonance tracking is to move a probe tone in frequency space rather than phase space, we require a linear conversion factor to change our phase feedback into frequency feedback. Luckily as seen in Fig. 2, resonators have a linear regime in the relationship between phase and frequency which can be measured quite simply.

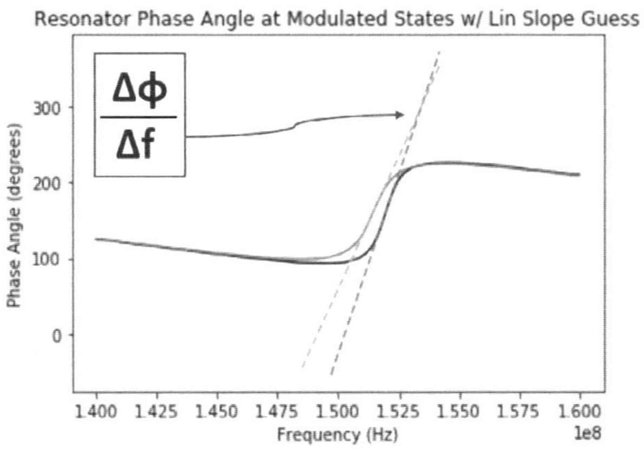

Fig. 2. A plot highlighting the linear regime in the phase/frequency relationship of a resonator in both the rest state and a modulated state. Previous applications of KIDs have confirmed that resonant frequency will not stray from this linear portion. While modulation of the KID causes changes in slope, the tone tracking system is designed in a way where these differences can be ignored.

With the tools of an invariant and a conversion factor in hand, it is now possible to design an algorithm to track the resonant tone of a KID. Before observing, one can determine the original resonant frequency of the KID (f_{res0}), the resonant phase (φ_0) and the linear regime slope ($\Delta\varphi/\Delta f$). Once power falls on a KID, there will be a change in phase of ($\Delta\varphi$). Putting these together, it only takes a simple set of equations to reach our final goal of the new resonant phase (f_{res_new}).

$$(\Delta f/\Delta\varphi) * \Delta\varphi = \Delta f \quad (1)$$

$$f\_res0 + \Delta f = f\_res\_new \quad (2)$$

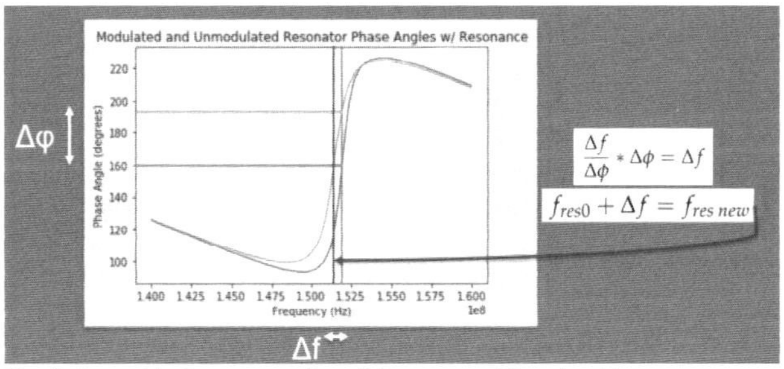

Fig. 3. A graphical representation of the tone tracking algorithm.

Fig. 3 provides a graphical representation of the tone-tracking algorithm and how it uses measurable quantities and known relationships to shift probe tones in response to changes in incident power on the detector. In digital control terms, the tone-tracking algorithm is a combination of a phase-locked loop (PLL) feedback-controlled with a proportional/integrative (PI) controller.

The block diagram shown in Fig. 4 shows in more detail the tone-tracking algorithm for a single detector. Starting with a probe tone at the KID's original resonant frequency, the signal is outputted by the digital to analog converter (DAC) and travels through the transmission line coupled to the KID. The affected tone then re-enters the FPGA by a digitizer (ADC) in quadrature and is multiplied with a complex conjugate of the probe tone to mix down to DC. The phase of the DC signal is calculated and fed through a filter to remove small fluctuations. A differential is taken between the new phase and the resonant phase to find the keystone value of $\Delta\varphi$. This differential is multiplied by the inverse slope of the estimated phase/frequency ratio to convert it to a differential frequency. Since there is lag time due to the length of the tone-tracker as well as uncertainty in the accuracy of the linear regime slope, this frequency difference (Δf) is scaled down considerably as to not "overshoot" the goal. The scaled down (Δf) is then added incrementally over multiple loops through the firmware to the original probe tone and inserted into a CORDIC wave generator to make a new probe tone closer to resonance. Finally, the probe tone exits the DAC and repeats the cycle until resonance is found and a phase differential of zero is achieved.

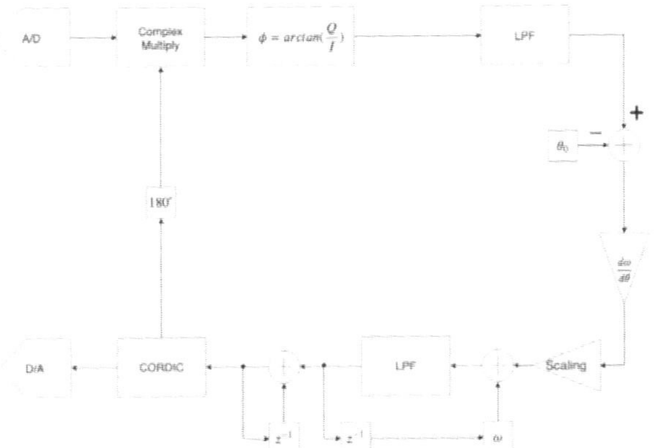

Fig. 4. A block diagram of the single-channel tone-tracking algorithm firmware. The diagram can be interpreted as follows: 1. Probe tone leaves DAC and goes through observing KID 2. Resulting tone returns to the ADC and is mixed to DC 3. Phase of input is calculated and integrated 4. Differential is taken to find $\Delta\varphi$ 5. $\Delta\varphi$ is scaled by estimated linear slope ratio 6. Differential frequency is added to original probe frequency 7. Probe tone is made with a CORDIC design and enters DAC 8. Return to step one and repeat

III. Simulation and Results

A firmware based on the algorithm explained in Fig. 4 was developed within the Simulink System Generator environment in MATLAB. The System Generator function of MATLAB's Simulink is the steppingstone between the block diagram and usable firmware as it allows for graphical DSP design and converts the designs into a hardware defined language (HDL) compatible with FPGAs. While the structure of the firmware is similar to the block diagram, the digitizer we plan to use operates at twice the clock speed of the FPGA fabric and therefore two samples will be fed for every clock cycle on the FPGA. To account for this, a parallel design was created so that the tone-tracker can handle two samples at once.

With proper handling, Simulink can also act as an extremely effective method of simulating firmware designs. To accomplish this, a sufficiently analogous "digital resonator" was developed within MATLAB and fed into the ports of the tone-tracking design. Simulations were executed starting at the simplest possible model, a simple resonance shift in a noiseless environment. Two shifts were used, one on the scale that is expected to be seen during observation and one significantly larger to stress test the algorithm. In the larger case, the algorithm found resonance on the order of 10 microsecond with a small overshoot. The smaller resonance shift was relocated in less than half the time with no oscillations. The time taken until a residual phase error of zero was reached can be seen in the plot in Fig. 5.

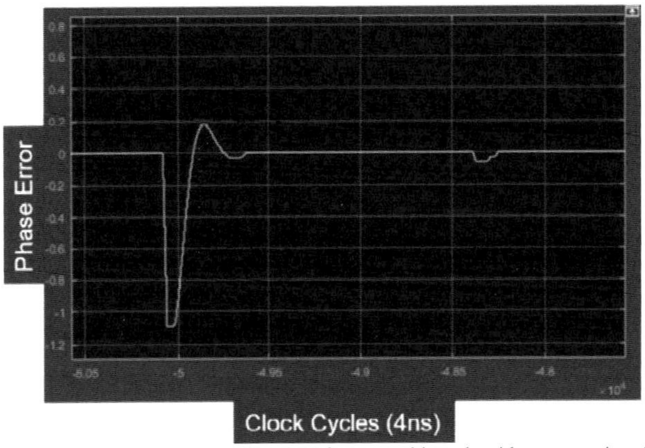

Fig. 5. Noiseless simulation results of tone-tracking algorithm measuring the difference between the real-time phase and resonant phase. Zero means the tone is tracked. The first spike represents a large phase shift and a dampened oscillation is visible. The second bump is a much smaller and more realistic phase shift.

For a more realistic scenario, white noise was added to the system and another set of tests was run, still using the same pair of large and small shifts. Fig. 6 shows these two simulated resonance shifts with an added white noise component. The tone tracking algorithm located the resonance in both cases on millisecond timescales with a negative-exponential-shaped curve.

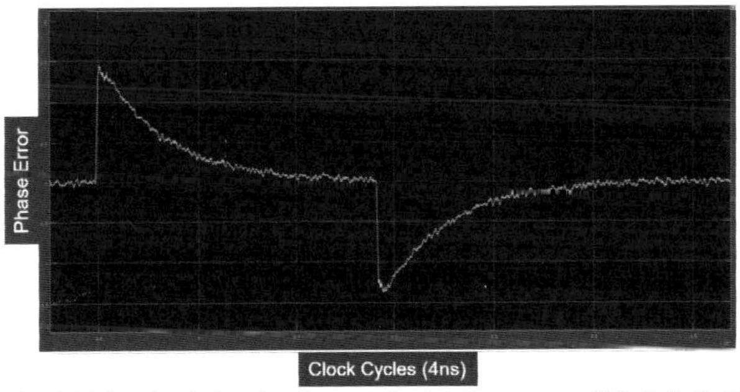

Fig. 6. Noisy simulation showing two simulated resonance shifts both finding the resonance on millisecond timescales.

IV. Discussion

The results of the tone-tracking simulations, while seemingly trivial, are a fantastic milestone in proving the validity of the algorithm for locating resonant frequencies in-situ. The noiseless examples seen in Fig. 5 verify that the fundamental algorithm used is an effective control system for reducing phase error to zero. Tweaks to the weight of the integrative portion in the P.I. controller could improve the speed and accuracy even further.

The noisy simulation in Fig. 6 is much closer to what will be encountered during observation. The results of this test imply that the time required to relocate resonance is largely independent of the modulation magnitude. The negative exponential shape of the phase differential implies that noise will significantly increase the timescale of relocating resonances. However, the shape of the plot also shows that the integrative portion of the controller is effective in eliminating the overshooting seen in the noiseless simulations. In its current state, the noisy tone-tracker was able to relocate resonances on millisecond timescales. Detector sampling rates from BLAST-TNG and other projects are on a similar millisecond timescale, confirming that the current tone-tracking algorithm is just fast enough to be effective in similar observation scenarios.

V. Conclusion

The ability to constantly track the resonant frequency of MKIDs in-situ has been desired for quite a while and the complexity of the problem speaks to why it still does not exist. Significant progress has been made over the past year in developing an FPGA-based tone-tracking system, but there is a great deal more to accomplish before it can be used in future MKID observations.

Since correctly tuning bit growth of the signal through the firmware seems to be posing a considerable hurdle, a Python simulation imitating the algorithm is under development that will focus on where bit-widths can be reduced. By using a software testing environment instead of firmware simulations, we can drastically reduce the iteration time for perfecting bit-propagation.

Furthermore, developing a firmware module which can read and update the tone-tracker in real time with the slope of the linear phase/frequency regime would greatly reduce the amount of time needed to relocate resonance. The current design for the tone-tracker can ignore the differences in these slopes by using very small iterative steps. By knowing the exact slopes, step size towards the new resonance can be maximized without overshooting and therefore reduce the number of clock cycles needed.

With the above steps completed, a single-channel version of a tone-tracking algorithm can finally be compiled and uploaded to an FPGA for tests on room temperature resonators and eventually cooled MKIDs. Once the single-channel design finds success, the firmware will be expanded to its full 1024/2048 channel form for use in MKID arrays. Finally, all designs, software, and firmware modules will be uploaded to GitHub and made open source for others to use and modify so that MKIDs can continue their rising popularity in the astronomy community

VI. References

1. Mauskopf, P. D., "Transition edge sensors and kinetic inductance detectors in astronomical instruments," Publications of the Astronomical Society of the Pacific 130, 082001 (Jun 2018).

Jonathan R. Hoh is a Ph.D candidate in astrophysics and a Master's student in Microwave engineering at Arizona State University, School of Earth and Space Exploration. His academic interests include the design and miniaturization of Terahertz frequency heterodyne detection systems along with the implementation and readout of microwave kinetic inductance detector.

Arizona State University Ultra High Gain Low Distortion Cryogenic Low Noise Amplifier

Justin L. Mathewson, *Arizona State University,* Jonathan R. Hoh, *Arizona State University,* and Hamdi Mani, *Arizona State University*

Abstract— **Arizona State University's radio astronomy lab has developed several cutting edge custom low-noise amplifiers (LNAs) over the past year. These LNAs cover a range of bandwidths and operate at room and cryogenic temperatures. Here we present one of the custom LNAs that is a multichannel, high-gain amplifier for use at baseband frequencies in terahertz heterodyne receivers. The LNA has wide a bandwidth of 0.1-3 GHz, low noise (NF < 1 dB) and an impressive gain of 90 dB. This LNA uses cost-effective, commercially available components and is packaged for easy integration into receivers in a custom compact enclosure that divides each channel into three isolated compartments to prevent crosstalk and oscillations. A combination of Silicon Germanium (SiGe) HBT and Gallium Arsenide (GaAs) HEMT transistors have been used to achieve the desired performance levels. A bias-dependent gain flatness of less than +/- 1 dB has been measured on the 0.5-3 GHz band. Power consumption per channel ranges from 195.5 mW at 90 dB of gain down to 71.5 mW at 80 dB of gain while maintaining state-of-the-art levels of power dissipation.**

Index Terms—**Baseband, Cryogenic, High Gain, LNA, Low Noise Amplifier, Wideband**

I. INTRODUCTION

For the past decade Arizona State University (ASU) has been on the cutting edge of custom made cryogenic low noise amplifiers (LNA). The standard LNA's built have a frequency range of 10MHz to 2GHz and 0.5GHz to 3GHz. Both yield a gain of 30dB at relatively low power consumption (300K: P_{2GHz} = ~40mW, 10K: P_{2GHz} = ~13mW; 300K: P_{3GHz} = ~70mW, 10K: P_{3GHz} = ~10mW). This LNA was a custom build, created for a project called Comets Observation & Mapping Terahertz Spectrometer (COMETS), a Terahertz Mapping spectrometer satellite. The project is led by NASA JPL [1]. The LNA uses commercially off the shelf components to achieve its high gain flatness.

II. MATERIALS & METHODS

Using commercially off the shelf components, from the standard LNAs that are built at ASU, such as Silicon Germanium (SiGe) Heterojunction Bipolar Transistor (HBT), Gallium Arsenide (GaAs) High Electron Mobility Transistor (HEMT) and gold wire bonds, were chosen to achieve these performance levels. A bandwidth of 0.1 to 3GHz and a noise factor of <1 that has a bias dependent gain flatness of ±1 dB. Each channel is separated into three compartments in a custom designed compact enclosure. Its noise temperature is ~80K at room temperature and can be easily integrated as an effective IF amplifier for terahertz receivers.

Fig 1. Shows the 8 channels and their individual and separated amplifier PCB's.

III. Results

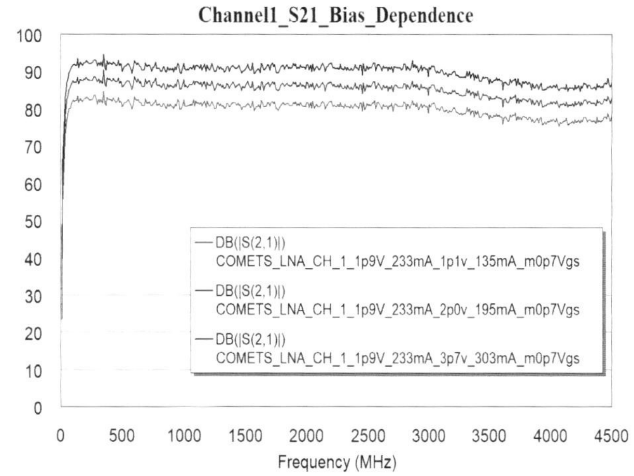

Fig. 2. S-parameters showing the bias dependency and the gain flatness for Ultra High Gain Low Distortion Cryogenic Low Noise Amplifier

Figure 1 plots the gain through the amplifier at three separate biases. At full bias (~195.5 mW per channel), a stable 90 dB gain was continuous across the entire bandwidth with a gain flatness of (+/-) 1 dB. The lowest bias tested (~71.5 mW per channel) still was able to maintain 80 dB of gain across the bandwidth with a similarly impressive gain flatness. The LNA can be biased between these power levels as well for intermediate gain.

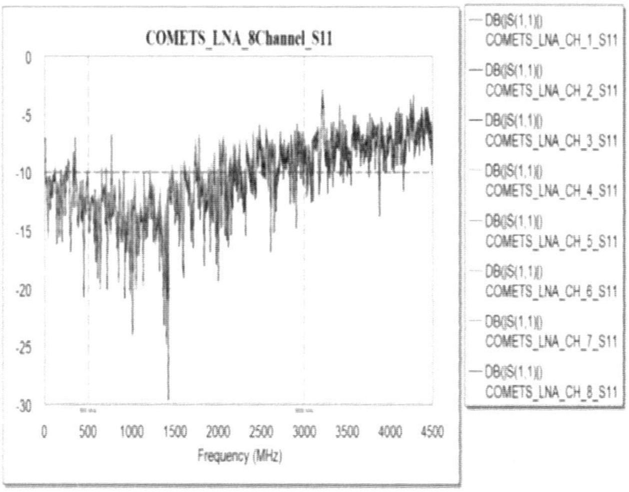

Fig. 3. A plot of the S11 vs Frequency. As Seen, S11 Is below -5 dB across the 0.5 to 3 GHz span.

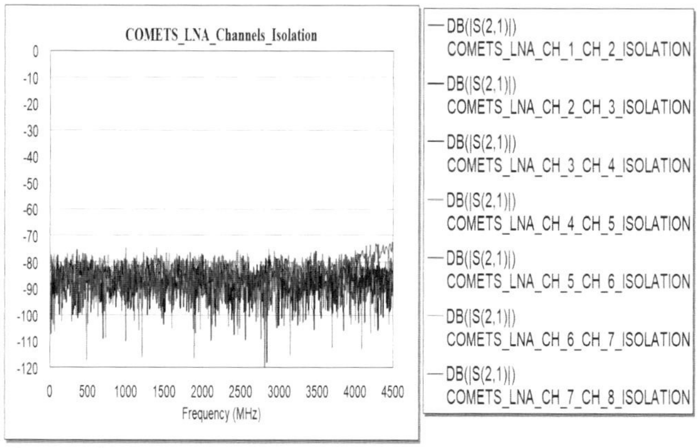

Fig. 4. Isolation across the 8 channels of the amplifier show that there is no crosstalk.

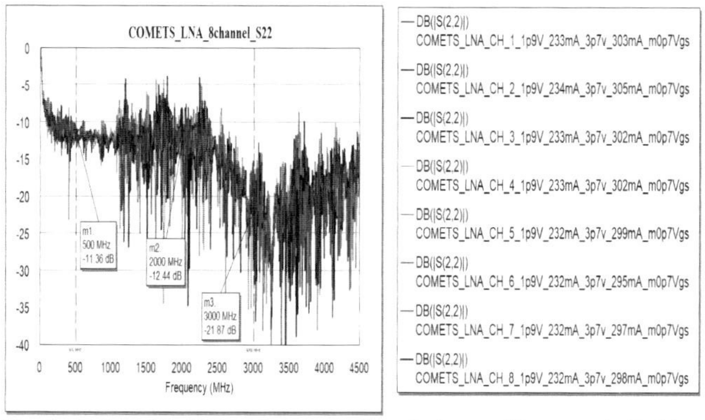

Fig. 5. A plot of the S22 vs Frequency. Seen, is the amplifier around -10 dB with partial spikes showing up barely exceeding -5 dB

Referring to figure 2, we see the gain (S21), with ±1 dB gain flatness. Figure 3, shows the amplifier's reflection (S11), which stays below -10 dB from 0 to about 2 GHz, but stays below -7 dB from 0 to 3 GHz. Figure 4, illustrates that there is negligible crosstalk between the 8 channels of the amplifier when all channels are powered at the same time. Last we have figure 5; showing the reflection at the tail end of the amplifier measurement (S22), which stays below -10 dB with a few spikes here and there that barely exceed -5 dB.

IV. Discussion/Conclusion

The data has come to fruition through days of design and testing using a 10K cryostat and off the shelf components. After significant designing, building, and testing, the JPL COMETS LNA has fulfilled all intended goals. What was achieved was a bias dependent amplifier whose frequency spans from 0.1 to 3 GHz and has a flatness of ±1 dB. It can achieve 90 dB of gain at relatively low power consumption. The minimum power that was tested was 71.5 mW yielding 80 dB of gain and the max at 195.5 mW yielding 90 dB of Gain. With a noise temperature tested at room temperature is at ~80K and the believed noise factor is <1.

Future endeavors for this amplifier, is to build several stand-alone single channel versions to show repeatability not only among the gain but also with the P1dB and IP3. To see the amplifier in use, a similar amplifier will be used for ASTHROS.

Justin L. Mathewson received a B.S. from Arizona State University from the School of Earth & Space Exploration in Exploration Systems Design in 2015. He works in the Radio Astronomy Terahertz laboratory as a Research Specialist at ASU. His main work is in designing, building and testing room and cryogenic low noise amplifiers for radio telescope receivers and projects that operate at different wavelengths.

Jonathan R. Hoh is a Ph.D candidate in astrophysics and a Master's student in Microwave engineering at Arizona State University, School of Earth and Space Exploration. His academic interests include the design and miniaturization of Terahertz frequency heterodyne detection systems along with the implementation and readout of microwave kinetic inductance detector.

Hamdi Mani worked in the electrical engineering department at Caltech as a research assistant from 2005 to 2010 developing cryogenically cooled low noise amplifiers for radio astronomy. He graduated from Arizona State University with a B.S in Physics and works in the Radio Astronomy Terhertz laboratory at ASU. He is an RF/Microwave engineer developing RF/Microwave instruments for an assortment of radio telescope receivers and projects that operate at different wavelengths. His main work and interest is in extremely low noise amplifiers (LNAs) operating at room and cryogenic temperatures along with noise measurement techniques for ultrasensitive LNAs.

Characterization of Cryogenic Flexible Transmission Lines Designed for the GUSTO IF Harness

Marko Neric, Hamdi Mani, Thomas Mozdzen, and Chris Groppi

Abstract—**Cryogenic flexible printed circuit ribbons based on stripline transmission lines have been custom designed, fabricated, and characterized for an upcoming NASA Class-D balloon mission. The Galactic/Extragalactic Ultra-Long Duration Balloon (ULDB) Spectroscopic Terahertz Observatory (GUSTO) will make use of 8-channel flex ribbons to transmit Intermediate Frequency (IF) from 0.3-4 GHz while at cryogenic temperatures and in flight. In this configuration one individual flexible ribbon can replace up to 8 semi-rigid SS-SS coaxial cables that are 20 inches in length. The GUSTO focal plane array consists of 24 pixels that would otherwise require an equal number of coaxial cables. Similar performance can be achieved using only 3 sets of flex ribbons, each of which is comprised of two circuits in series. This will not only reduce the weight and form factor of the IF harness, but also provide improved thermal performance. The ground planes of the flex circuit have been reduced to minimize heat transfer into a cryogenic system which can extend mission lifetime by mitigating coolant boil off. Each channel of the flex circuit conducts a quarter of the heat that a coaxial cable would. To accommodate the unique design of the GUSTO cryostat three pairs of flex ribbons will be RF coupled to provide the necessary length and shape for routing between IF components. Flex circuits were designed with a staggered array of SMP type press on connectors instead of SMA which makes coupling 8 channels achievable in one easy maneuver. The circuits can be clamped together to prevent separation in flight or other high vibration conditions all in less time than it would take to install coax. All RF testing was done using a closed cycle vacuum cryogenic station with the test bed held at 20 K. Each pair of flex ribbons was thermally coupled to the 20 K stage. The flexible transmission lines showed an average insertion loss of 3.07 dB/ft at 5 GHz, while the industry standard SS-SS UT-85 coax loss is 2.80 dB/ft. The GUSTO IF system will use state of the art ASU cryogenic low noise amplifiers to more than make up for the higher insertion loss. Any future systems that can correct for, or accept the additional loss will benefit more from the other features of the flex circuit design. The custom made cryogenic flexible transmission lines are still the preferred method of signal transmission for applications that require complex routing, lightweight components with small form factor, easy assembly, lower heat transfer, and flexibility.**

Index Terms—**RF, Cryogenics, Transmission Lines.**

I. Introduction

The intermediate frequency (IF) harness of the Galactic/Extragalactic Ultra-Long Duration Balloon Spectroscopic Terahertz Observatory (GUSTO)[1] has been designed, fabricated, and characterized. GUSTO is a NASA balloon mission that will observe THz signals from within the Milky Way and the Large Magellanic Cloud. GUSTO will have a 24 pixel focal plane made up of three separate 8 pixel arrays. Each array is housed inside the same vacuum cryogenic system. The IF harness must transmit the signals of 24 independent pixels from within the cryogenic station to an outside interface.

Two choices for a harness were considered. The first was 24 sets of stainless steel - stainless steel (SS-SS) semi rigid coaxial cables. The benefits of semi-rigid coax include their well characterized performance, and their repeated use in commercial and space applications which gives coax a long heritage. Stainless steel is often chosen for transmission in and out of vacuum cryogenic systems due to its low thermal conductivity despite more signal loss than alternatives. The drawbacks to using coax become more evident as the pixel count in an instrument increases. The resulting increase in pixels requires an increase in transmission lines. For a system using coax this means each cable must be individually installed to the same specifications. This increases the risk of individual failure while also increasing the overall form factor of the harness and installation time. The industry standard SS-SS semi rigid coax is UT-085 which has an outer diameter of 0.087". Smaller cables than this are available but tend to be more fragile and higher in loss. Semi-rigid cables can be more difficult to route through a cryogenic system since they cannot accommodate paths with sharp turns or some narrow spacing. It often becomes necessary to install hermetic vacuum feed-through adaptors at several points within a cryostat to complete the harness which further increases the number of coax lines and potential areas of failure.

The second option considered was to make the IF harness using ASU cryogenic flexible transmission line ribbons[2]. The ASU flex ribbons use a stripline design as a base and can incorporate several channels in a single ribbon. A prototype 8-channel ribbon was designed previously with the GUSTO mission in mind and to test key features that would give the flex ribbons an edge over coax. The prototypes were 8-channel circuits in either a straight ribbon or a curved ribbon to test the effects of in-plane turns on the individual channels. Each prototype had a width of 1.1", and a total length of 20". The ribbons were populated with press on connectors that allow fast an easy connections. They can be custom designed to meet the specific routing needs of the mission without significant signal loss. The prototypes achieved similar RF performance to coax, and transfer significantly less heat into the cryogenic system which has a limited supply of coolant. The drawbacks to this approach are that the best RF performance is achieved when the ribbon is at cryogenic temperatures, and that is still slightly higher in signal loss than SS-SS coax. Low noise amplifiers can be used to boost the IF signals to make up

M. Neric, H. Mani, T. Mozdzen, and C. Groppi are with the School of Earth and Space Exploration, Arizona State University, Tempe, AZ, 85287 USA e-mail: mneric@asu.edu.

for the slightly higher loss of the flex ribbons. The custom ribbons can meet the unique routing needs of the GUSTO mission without requiring alterations to the cryostat itself. For the above reasons the decision moving forward was to build a harness out of flexible ribbons.

II. DESIGN

The limiting size of the individual flexible ribbons is the size of the panel which the flexible printed circuit material is made from. These have a working area of about 18" x 20". To mitigate cost it is advantageous to fit as many flexible ribbons on the same panel as possible. The average length of GUSTO's IF bands is 3.30 feet and each band takes a different path inside the cryostat. Due to these constraints a single ribbon could not comprise the entire harness. Instead each of the three IF bands is made of two flex ribbons. One ribbon designated a "cold" circuit because it is held closest to the 20 K stage of the cryostat, and the other known as the "warm" circuit which would interface with the outside of the cryostat. The cryostat designed for the GUSTO mission had a particularly challenging path the IF harness needed to go through. The flexible transmission lines were custom designed with short in-plane turn radii (inside radius no less than 1"). This resulted in 6 uniquely shaped ribbons. Figure 1 shows the GUSTO cryostat and the installed IF bands. Figure 2 shows the reference design for one of the "cold" circuits.

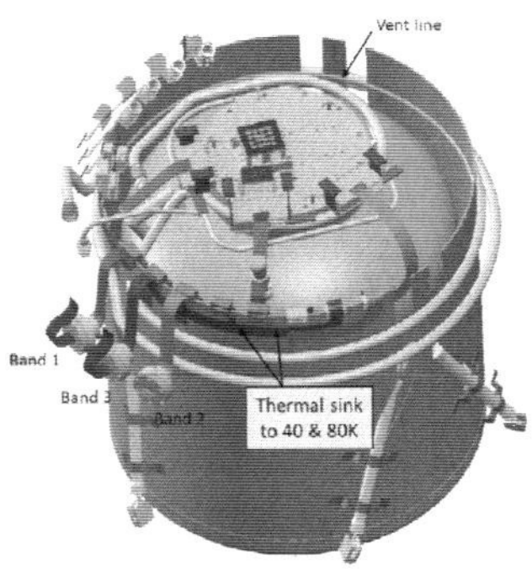

Fig. 1. This image of the GUSTO cryostat demonstrates how complex the routing of the IF harness will be. The custom designed flex circuits are shown as bands 1-3. Flexible transmission lines simplify the install process since the ribbons snap together to complete 8 signal paths whereas the same system would have needed a minimum of 24 discrete semi rigid coaxial cables for the task.

For each band the ribbons required two points of contact with the cryostat. This is made possible by small 0.50" x 0.50" copper heat-sinking tabs that can be directly soldered to. The operating temperature for the GUSTO mission is a 20 K - 300 K gradient and using both heat-sinking tabs helps maintain this. Each ribbon is an eight-channel transmission line with the connections staggered so they can connect to one another. DELTA Electronics smooth-bore SMP press-on connectors were chosen because of their small form factor and low retaining forces. Limited or full detent versions may also be used, however their higher mate/de-mate force may reduce the lifetime of a ribbon after repeated use. The eight smooth-bore connections are sufficient to hold the flex circuits

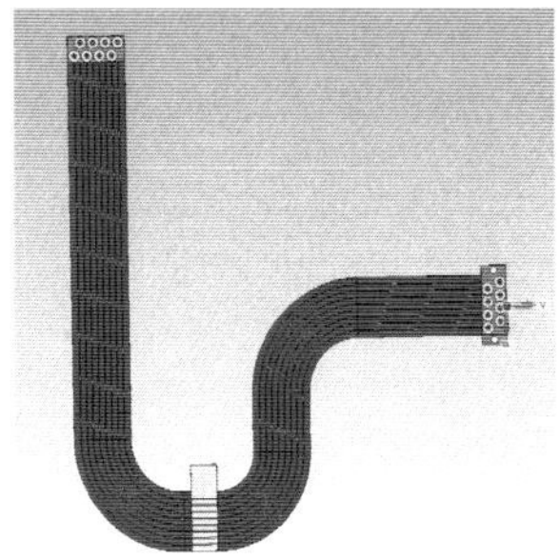

Fig. 2. Drawing of an individual custom flex circuit cable. There are two per band, and three bands total within the IF harness. Each circuit has a unique shape in order to facilitate the complex routing within the GUSTO cryostat. Visible in this drawing is a small heat-sinking tab that can be directly soldered to. Each band has two tabs that can be attached to separate stages of the cryostat for temperature control. Also visible are fastener holes used to secure the circuit on one end. Copper is shown in yellow, the kapton substrate is in red, and the top layer of soldermask is shown in grey. The yellow copper stripes that follow the path of the substrate are copper ground planes necesary for stripline transmision. These are not the signal traces themselves. The signal traces are burried under the substrate layer.

together, however a clamp may be used for added security. Fastener tabs like the one shown in figure 2 may be used for this purpose.

The copper grounding planes on top and bottom of the flex circuit are required by the stripline. The overall metal content on both sides of the ribbon has been reduced to narrow 0.015" strips. This pattern lets the individual channels behave as a stripline and reduces overall heat transfer in and out of the system. Copper has a high thermal conductivity of 401 $WK^{-1}m^{-1}$ compared to that of stainless-steel which is 16.3 $WK^{-1}m^{-1}$. Reducing the metal content of the flex ribbon drastically changes the cross sectional area of conducting metal. The result is that a single channel of the flex ribbon conducts 25% of the heat a UT-085 SS coax cable would.

The ribbons can be bonded to custom made vacuum flanges to break out of the cryostat. Prototype circuits were potted directly to modified Aluminum KF flanges using commercial scotch weld epoxy adhesive. The adhesive can be applied directly to to the circuit. As mentioned above and as shown in Figure 2, there is a reduced copper ground plane on top and bottom of the flex ribbon. This exposes the layer of kapton underneath. Bonding the ribbon to a vacuum flange over the substrate has been tested and shows no degraded performance. The aluminum flanges were helium leak checked to test their vacuum. No leaks were detected. The flanges were then thermally cycled, and leaked tested again with no leaks detected.

The flex ribbons designed for the GUSTO mission had a solid copper plane where the vacuum flange was to be potted. The vacuum flange was to be a modified stainless-steel CF flange commonly used in high vacuum instruments. The first set of "warm" circuits were bonded to the CF flanges and vacuum tested. No leaks were present at the time of the initial leak check. The circuits were than thermally cycled for a temperature range of -40 C to +60 C for a total of six cycles. The bonded circuits were then leak tested again to make sure the adhesive held together. This time unfortunately two out of the three circuits registered an air leak above tolerance. The leak occurred at the interface between flange and adhesive.

A second set of modified CF flanges were made and this time the surfaces were prepared differently before bonding. The flange was bead-blasted to give it an abrasive surface that the scotch weld might better adhere to. Industrial grade chemical solvents were used to clean the CF flange of any machine oils or other contaminants that would prevent complete adhesion. The CF flanges were also cleaned using mineral spirits and isopropyl alcohol in a sonic bath. Test circuits were bonded to the cleaned flanges. Initial leak testing found no leaks. The flanges were thermal cycled and tested again, just as before, two out of three flanges were observed to leak. The leak was again detected on the interface between flange and adhesive. The flex circuit itself is not believed to be the source of the leak.

It is unclear whether or not the leak is a result of the thermal cycling method (i.e. time dependence of thermal cycles) or if the chosen adhesive does not retain its bond with stainless-steel surfaces due to a mismatch in expansion coefficients. Aluminum flanges have not had any leaks. Due to costs of aluminum CF flanges to replace the stainless-steel ones for GUSTO, and the time constraints of the mission, an alternate break-out scheme was designed. A CF flange with 8 hermetic feed through adapters will be used as the outside interface. The flex ribbons will reach this flange and a series of short coax lines will serve as the interconnects between flange and ribbon.

The boards were designed at ASU and fabricated by Coast-To-Coast circuits. The SMP press-on connectors were assembled and DC tested at GMA Manufacturing. Cryogenic and RF testing was done at ASU.

III. Results

A closed cycle vacuum cryogenic 10 K test bed was used to characterize the IF harness. All RF measurements were done using a Rhode & Schwarz ZVA 24 Vector Network Analyzer (VNA) that was certified calibrated and following a pre-approved test plan. The VNA test port power was set to 0 dBm, with 10x averaging and an IF frequency of 100 Hz. The measurement bandwidth was 0.1 - 6 GHz which encompasses the 0.3 - 4 GHz operating frequencies the IF harness will see during its mission. Due to the limiting size of the test bed only one band could be tested at a time.

The cryostat interface was a custom made bulkhead plate with four hermetic SMA feed-through adapters. Short mini circuits SMA-SMP UT-47 coax cables were used to transition from the bulkhead to the flex ribbon. Earlier tests were done using small flex ribbon segmants specifically made to serve as adapters for the flex circuit interface. The change to coax was done to better reflect the in flight conditions. These coaxial adapters were not heat-sunk and therefor closer to the ambient temperature outside the cryostat. In order to de-embed their effects from the flex ribbon room temperature measurement of the mini-circuits cables were recorded. The room temperature data was then subtracted from the cryogenic result at all frequency points. Four total bulkhead adapters were used in each measurement so all eight channels of the IF harness could be reached in a single cool-down. Measuring all eight channels at once saves time by reducing the number of cryogenic cycles per band. This also means that the environmental conditions inside the test bed remained unchanged for each ribbon.

The two flex ribbons of each band were fastened together using a clamp and then installed on the 10 K stage of the cryogenic test bed. One heat-sinking tab was thermally coupled to the 10 K stage. There was insufficient space to couple the second tab of any band and still be able to connect the flex ribbon to the cryostat output. This resulted in a much higher temperature gradient through the flex ribbon than the likely temperature profile it will have in flight. Thus all results presented are worse than expected for the actual mission. When the ribbon is fully installed a temperature sensor is placed in contact with the circuit at the heatsinking tab. Two resistor heaters were attached to the 10 K stage.

The test bed can reach $\sim$ 10 K temperatures in just under three hours. The system was run until a stable 10 K temperature was achieved, and the heaters were slowly dialed up so the cold plate temperature would stay above 20 K as is expected during the GUSTO mission. RF measurements were taken to test the signal transmission, reflections, and isolation or cross-talk. After a band was measured the heaters, and cryogenic station were turned off. The system was brought back to ambient room temperatures overnight without the use of heaters. Figure 3 is a plot of the average s-parameters of all three bands with an average band length of 3.3 feet.

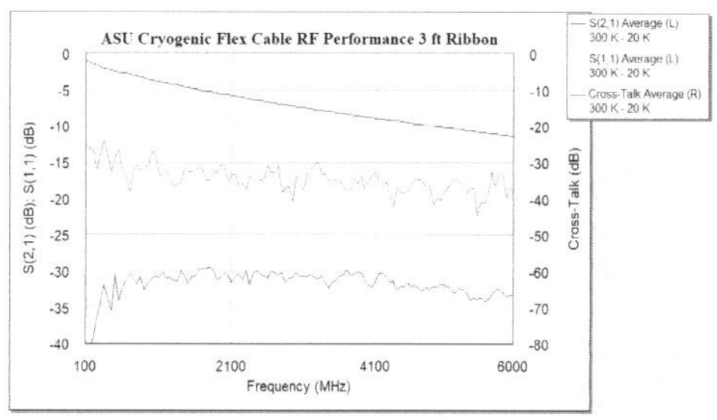

Fig. 3. Average s-parameters of all three IF bands are shown. The cryogenic testing was done with one end of the band held near 300 K and the other end held at 20 K. The average length of each band is 3.3 ft. At 5 GHz frequency the average loss per band is 3.07 dB/ft which exceeds design specifications. The cross-talk between adjacent channels is below -60 dB for nearly the whole bandwidth which also exceeds the specifications. There are no requirements for the reflection, however a value below -15 dB is usually desirable in RF applications.

The design specifications for the GUSTO IF harness were: IF bandwidth of 0.3 - 4 GHz, a signal loss no greater than 6.00 dB per 20 " length of harness at 5 GHz frequency, and 30 dB or more of isolation between channels. As shown in figure 3, the average loss of all three bands is 5.12 dB per 20" length. The measured cross-talk shows greater than 60 dB of isolation between channels for nearly the entire IF bandwidth. The circuits were measured from 0.1 - 6 GHz. Having met all specs the IF harness has been delivered for integration with remaining flight hardware.

IV. Conclusion

The GUSTO mission will showcase several of the best features and abilities of the ASU cryogenic flex ribbons. Upcoming instruments that have a high pixel count, complicated or limited routing space, and have need of compact and lightweight solutions for their IF system will benefit from using the ASU flex ribbon. The lessons learned from the GUSTO mission will drive further innovation of the flex ribbon that already offers comparable performance to industry standard semi-rigid coax with other measurable benefits. The ASU flex ribbon exceeds all design specifications that were set for the GUSTO mission. They achieve a bandwidth that includes 0.3 - 4 GHz with an insertion loss at cryogenic temperatures that is 3.07 dB/ft and Isolation much greater than 30 dB over the entire bandwidth.

Acknowledgment

The authors would like to thank Bob Giambrone, and Hector Ramirez of Coast to Coast circuits for their help and support during the fabrication of the flexible test circuits. The authors would also like to thank David Good, Gary Abramczyk, and Suzette Martinez of GMA Manufacturing for their help and support with the board assembly.

References

[1] C. Walker, C. Kulesa, P. Goldsmith, C. Groppi, F. Helmich, D. Hollenbach, J. Kawamura, W. Langer, G. Melnick, D. Neufeld, J. Pineda, G. Stacey, A. Stark, A. Tielens, M. Wolfire, H. Yorke, and E. Young, "GUSTO: Gal/Xgal U/LDB Spectroscopic-Stratospheric TeraHertz Observatory," Amer. ASTRO. Soc., AAS, 231, 2018

[2] P. McGarey, H. Mani, C. Wheeler, C. Groppi, "A 16-channel flex circuit for cryogenic microwave signal transmissions," SPIE, 9153, 2014

Marko Neric graduated from Arizona State University (ASU) in 2014 earning a B.S. in Physics. He is currently a Ph.D. candidate in the School of Earth and Space Exploration (SESE) at ASU. He is pursuing his Ph.D. in astrophysics, and an M.S. in electrical engineering. His focus is on THz astronomy, and instrument building.

Hamdi Mani has been an engineer at the ASU radio astronomy lab for the past 8 years. His main areas of research: design of wideband cryogenic low noise RF/ Microwave amplifiers and receivers, development of microwave and millimeter wave instrumentation, electronics packaging and cryogenic systems design . Prior to joining ASU, he worked at Caltech/JPL on developing SiGe cryogenic low noise amplifiers and wideband radio astronomy receivers.

Thomas J. Mozdzen received the B.S. in Physics in 1978 and the M.S. in Electrical Engineering in 1980 from the University of Illinois Urbana-Champaign, the M.S. in Physics from the University of Texas at Dallas in 1985, and the PhD in Astrophysics from Arizona State University in 2017. He has worked in the semiconductor industry for several years beginning as a reliability engineer at Mostek Corp, Dallas, Tx., and then mainly as a circuit design engineer at Siemens, Munich Germany, and at Intel, Chandler, Arizona. His PhD. Thesis focused on reducing systematic error in experiments to detect the sky-averaged redshifted 21 cm epoch of reionization signature. He holds several engineering related US patents and has authored several papers in both electrical engineering and astrophysics. He is currently employed at Arizona State University as a Postdoctoral researcher.

Christopher E. Groppi received the B.A. degree in astronomy (with honors) from Cornell University in Ithaca, NY, in 1997 and the Ph.D. degree in astronomy with a minor in electrical and computer engineering from the University of Arizona in Tucson, AZ, in 2003. In 2003, he joined the National Radio Astronomy Observatory as a Director's Postdoctoral Fellow. He then moved to the University of Arizona as an assistant staff astronomer in 2005. In 2006, he received an Astronomy and Astrophysics Postdoctoral Fellowship from the National Science Foundation. In 2009, he joined the Arizona State University School of Earth and Space Exploration in Tempe, AZ as an assistant professor. He became an associate professor in 2015. He is an experimental astrophysicist interested in the process of star and planet formation and the evolution and structure of the interstellar medium. His current research focuses on the design and construction of state of the art terahertz receiver systems optimized to detect the light emitted by molecules and atoms in molecular clouds, the birthplace of stars. He also applies terahertz technology developed for astrophysics to a wide range of other applications including Earth and planetary science remote sensing, hazardous materials detection and applied physics.

4 THz beam filter based on a back to back Si-lens system

Y. Gan[1,2], B. Mirzaei[1,3], S. van der Poel[3], J. R. Silva[1,2], M. Finkel[1], M. Eggens[1] , M. Ridder[1], A. Khalatpour[4], Q. Hu[4], F.F.S. van der Tak[1,2], and J. R. Gao[1,3]

We report a novel, spatial beam filter using two elliptical silicon (Si) lenses, mounted back to back, between which there is an opening aperture, to produce a Gaussian beam at 4 THz. The two Si lenses with a diameter of 10 mm have an antireflection coating layer, while the opening aperture is a circular hole on a thin Au layer defined on a double-side polished Si wafer. A mechanical lens holder is introduced to allow for aligning the 1st lens, the Si chip (with the aperture), and the 2nd lens as an integrated beam filter with an accuracy of 1 μm along their optical axis.

We demonstrate experimentally that such a back-to-back Si lens beam filter works using a 3.8 THz quantum cascade laser (QCL) as the source. We have shown that :

a) When the back-to-back Si lens has no aperture, namely only a bare Si wafer sitting between the two lenses, we obtained the transmission of a Gaussian beam through the system to be 72.5%±1%, which agrees to the simulation (77%) by COMSOL Multiphysics.
b) With an input beam of non-Gaussian profile (see Fig.1), and with an aperture, the back-to-back lens system shows an output beam with a Gaussicity of >98% and a transmission of 35-50 % from the input beam. The results are illustrated in Figure 1.

We are currently adjusting the aperture sizes and also input beams to optimize the performance of the beam filter. The ultimate goal of this research is to realize an ideal LO source for heterodyne arrays at supra-THz frequencies by combining a spatial beam filter with a QCL. An LO with an ideal Gaussian beam is crucial for optical system designs and for reducing stray light in practice.

A back-to-back Si lens beam filter is an alternative to the back-to-back (corrugated) metallic feedhorn approaches [1], which are known to be extremely difficult to be fabricated at the THz frequencies and also expensive. Furthermore, the latter has also a narrow bandwidth.

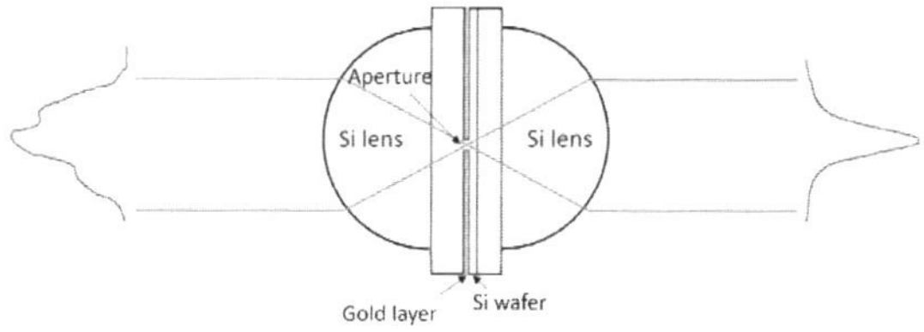

(a)

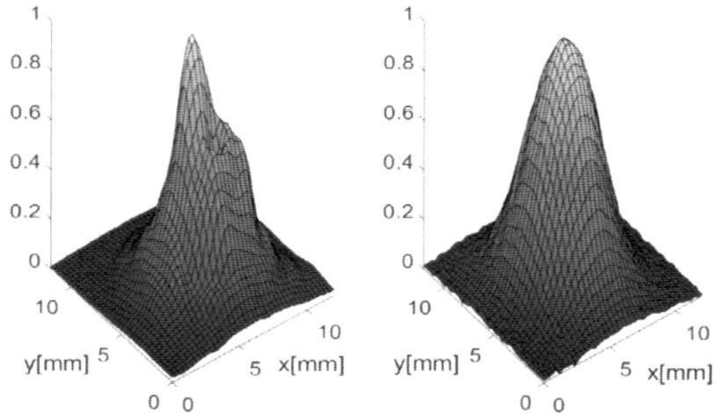

(b) (c)

Fig. 1. (a) The schematic of the back to back lens system. Two Si lenses are used to focus and collimate the incoming beam, and an opening aperture defined by the gold layer on the Si wafer is used to filter non-Gaussian components; (b) The 3D profile of the input beam, generated by a 3.8 THz QCL and collimated by a Polyethylene lens; (c) The 3D profile of the output beam from the back-to-back lens and aperture system. Compared to the input beam, the output beam has a nearly perfect Gaussian profile, with 99.8 % Gaussicity.

References

[1] As an example, see Hong Pu, et al., "A Broadband Back-to-Back Corrugated Horn Structure for Gaussian Mode Filtering in Terahertz Band," 2018 International Conference on Microwave and Millimeter Wave Technology (ICMMT), 18321784.

[1] SRON Netherlands Institute for Space Research, the Netherlands.
[2] Kapteyn Astronomical Institute, University of Groningen, 9747 AD, Groningen, The Netherlands.
[3] Faculty of Applied Science, Delft University of Technology, the Netherlands.
[4] Department of Electrical Engineering and Computer Science, Massachusetts Institute of Technology, USA.

NOTES:

Silicon Micromachined Receiver Calibration Waveguide Switch for THz Frequencies

Adrian Gomez-Torrent, Umer Shah, and Joachim Oberhammer

Radiofrequency (RF) switches are commonly used for the calibration of radioastronomy microwave instruments [1]. Optical systems, based on moving mirrors, are generally preferred when approaching the THz frequency range, due to the challenges to fabricate waveguide-based switches at such high frequencies.

One of the main challenges for the fabrication of THz waveguide components is their reduced size, in the micrometer range, which makes most traditional machining techniques highly unreliable for such applications. Silicon micromachining has been demonstrated to be a promising alternative for the fabrication of THz waveguide components [2], [3].

The authors recently demonstrated a silicon micromachined turnstile orthomode transducer [2] that could replace the bulky wire-grid polarizers currently used above the sub-millimeter wave range in many receiver architectures [1]. Here, we present a waveguide-integrated single pole double throw (SPDT) microelectromechanical system (MEMS) switch. This novel switch concept is based on a reconfigurable MEMS surface [3] that separates a stack of two hybrid couplers, see Fig. 1a, b.

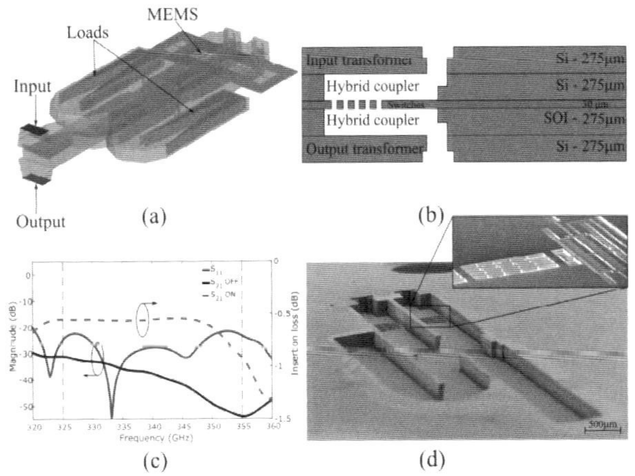

Fig. 1. On/load switch concept showing: (a) 3D drawing of the structure, (b) cross-section of the micromachined wafers (not to scale), (c) simulated performance, and (d) SEM image of the fabricated prototype.

The SPDT switch can be used for receiver calibration by loading one of the outputs of the switch with an integrated micromachined absorber. This configuration provides the ability to switch between an 'ON' state and a 'LOAD' state in the receiver. The simulated RF performance for both states is depicted in Fig. 1c.

The design of the switch is based on a stack of four micromachined chips. A cross-section of the chip stack is shown in Fig. 1b. The top and bottom chips include stepped impedance transformers, and E-plane bends to feed the in-plane micromachined waveguides. The signal is then split by an H-plane hybrid coupler that feeds the two MEMS reconfigurable surfaces. When the surfaces are in the open state, the signal is routed to the lower chip and recombined by the second coupler into the output port; if the surfaces are in the closed state, the signal is reflected and recombined in the loaded port. More information about the MEMS reconfigurable surface can be found in [3].

The fabrication of the SPDT switch is based on the multistep deep reactive ion etching (DRIE) technique described in [2], and on the waveguide-integrated MEMS technology described in [3]. Fig. 1d shows a scanning electron microscope (SEM) image of a fabricated prototype. After micromachining, the four silicon chips are gold-sputtered, and thermal compression bonded.

The SPDT switch in the ON/LOAD configuration, designed for 340 GHz, shows a simulated insertion loss in the 'ON' state of 0.5 dB with a return loss better than 20dB. In the 'LOAD' state, the predicted return loss is better than 20dB with isolation better than 30dB. A more complex architecture with several calibration loads could also be implemented by integrating additional cascaded switches in the same chip.

The authors are with the Division of Micro and Nanosystems, School of Electrical Engineering and Computer Science, KTH Royal Institute of Technology, Stockholm SE-100 44, Sweden. Email: adriango@kth.se.

References

[1] M. C. Carter et al., "ALMA front-end optics," presented at the SPIE Astronomical Telescopes + Instrumentation, USA, 2004, p. 1074.

[2] A. Gomez-Torrent, et al., "Compact silicon-micromachined wideband 220–330-GHz turnstile orthomode transducer," IEEE Trans. THz Sci. Technol., vol. 9, no. 1, pp. 38–46, Jan. 2019.

[3] U. Shah et al., "A 500–750 GHz RF MEMS Waveguide Switch," IEEE Trans. THz Sci. Technol., vol. 7, no. 3, pp. 326–334, May 2017.

Characterization of Dielectric Material at 300 GHz for Vacuum Window Applications

Keara Carter[1], C. Edward Tong[1], and Jake Connors[2]

Abstract— **This paper describes the experiments that have been used to characterize dielectric samples applicable for a low-loss vacuum window at 300 GHz. The complex refractive indices of the samples were measured with a Quasi-Optical Vector Network Analyzer (QO-VNA). The loss tangents were further confirmed with the Method of Intersecting Lines using a Superconductor-Insulator-Superconductor (SIS) receiver. Finally, leak rates of the samples were compared using long term monitoring on a small vacuum chamber.**

Index Terms—**Dielectric materials, dielectric measurements, superconductor-insulator-superconductor (SIS) receivers, sub-millimeter wave technology, vacuum window.**

I. Introduction

VACUUM windows form an important component for ultra-low noise submillimeter cryogenic receivers. Recent developments in receiver technology have driven the quest for wide aperture, low-loss, broadband, and reliable vacuum windows. Although there exists a wealth of information in the literature on the electrical and mechanical properties of different dielectric materials, the aim of our work is to explore new material and to confirm the properties of material from different vendors.

Future wideband Submillimeter Array (wSMA) receivers [1] require windows with a clear aperture diameter of 105 mm, covering an octave frequency range centered at 300 GHz. Large format receivers for Cosmic Microwave Background (CMB) observations necessitate windows with much larger diameter [2]. Quartz windows in use for the SMA receivers and ALMA offer only narrow band coverage and are somewhat smaller in size. That is why our focus is on thin but relatively strong window material that can be coated with an effective anti-reflection (AR) coating.

II. Quasi-Optical Vector Network Analyzer

We have set up a Quasi-Optical Vector Analyzer (QO-VNA), operating between 210 and 370 GHz, to measure the complex refractive indices of different materials. Fig. 1 shows the QO-VNA set up. The material under test is placed on the sample holder, attached to a translation stage located between two 90-degree off-axis parabolic mirrors with focal lengths of 50.8 mm. Efforts are made to ensure the sample is as flat and perpendicular to the incoming wave as possible. A bi-directional coupler on the source side uses harmonic mixers to monitor the amplitude and phase of the incident wave as well as the reflected wave for reflection measurements. The transmitted wave is measured by another harmonic mixer connected to the receiving horn. To reduce back reflections from the receiving horn and harmonic mixer, a reflective neutral density filter, which acts as an attenuator (of -12.5 dB at 300 GHz), is placed between the sample holder and the receiving mirror at a roughly 45-degree angle. Fundamental Gaussian analysis of this setup shows that the beam waist, where the wave front is planar, is located very close to the sample holder. The value of the beam waist radius is about 12 mm.

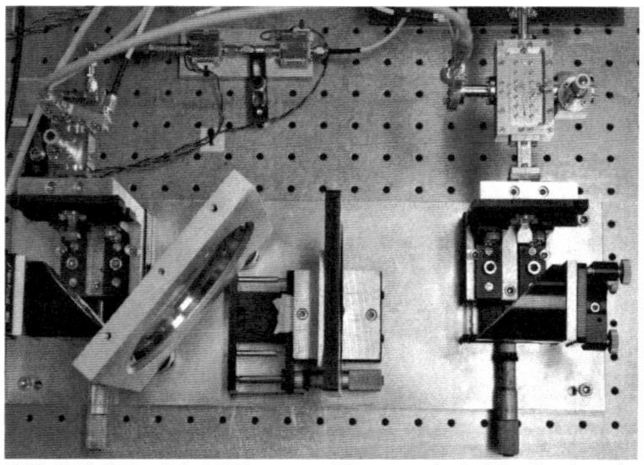

Fig. 1. 210-370 GHz QO-VNA setup. The sample holder is placed midway between a pair of off-axis parabolic mirrors. A partially reflective mesh, placed at 45° to the optics path, is put behind the sample holder for transmission measurements.

In order to derive the refractive index, a pair of transmission measurements are made using the QO-VNA: one with no sample and one with the sample under test mounted in the sample holder. The complex ratio of these two vector measurements, R, describes the geometric series of multiple reflections and loss, and can be written as follows [3]:

$$R = \frac{S_{21}^{sample}}{S_{21}^{thru}} = \frac{(1-\Gamma^2)e^{-j(n-1)\beta_0 d}}{1-\Gamma^2 e^{-j2n\beta_0 d}} \quad (1)$$

In the above equation, Γ is the reflection coefficient at the interface between air and the sample, the reflective index of

[1]K. Carter and C. E. Tong are with Center for Astrophysics | Harvard & Smithsonian, Cambridge, MA 02138, USA. (Email: kearaj.carter@gmail.com)
[2]J. Connors is with National Institute of Standards & Technology, CO 80305, USA.

which is n. n is a complex quantity, and its imaginary part is related to the loss tangent ($\tan\delta$) of the material.

$$n = n' - jn'' = n'(1 - j\frac{\tan\delta}{2}) \quad (2)$$

In a first order approximation, the phase of the complex ratio R varies linearly with frequency, and n' can be derived from the slope. The imaginary part of n can also be extracted from a plot of the magnitude of R vs. frequency, provided that the sample is thick enough to present measurable losses for the QO-VNA, the threshold of which is 0.1 – 0.2 dB. Once an approximate value is established for n' and n'', a final round of complex data fitting yields the final experimental value of n together with an estimation of experimental errors. The measured refractive indices and loss tangents of selected materials are listed in Table 1.

TABLE 1
REFRACTIVE INDICES AND LOSS TANGENTS OF SELECTED MATERIALS MEASURED WITH A QO-VNA

Material	Refractive Index, n	$\tan\partial \times 10^{-3}$
High Density Polyethylene (HDPE)	1.5327 ± 0.0005	0.6 ± 0.1
Honeywell Spectra (Dyneema)	1.552 ± 0.001	1.4 ± 0.3
Polypropylene (PP)	1.502 ± 0.001	0.5 ± 0.3
Polystyrene (PS)	1.594 ± 0.007	3 ± 1
Polytetrafluorethylene (PTFE)	1.434 ± 0.001	
PREPERM 225	1.605 ± 0.005	6 ± 1
Polymethylpentene (TPX)	1.45 ± 0.01	
Ultra-High Molecular Wt. Polyethylene (UHMW-PE)	1.515 ± 0.002	
Z- Cut Crystalline Quartz	2.107 ± 0.002	
Zitex G-108	1.22 ± 0.02	

III. METHOD OF INTERSECTING LINES

The loss tangent of selected materials measured with the QO-VNA have been confirmed with an *in-situ* loss measurement using the Method of Intersecting Lines with an SIS receiver. This method was first proposed by Blundell [4]. Using the procedure outlined in [5], Y-factor measurements are performed using hot and cold loads for a range of Local Oscillator (LO) levels. From this, the equivalent input noise temperature of the RF input section of the receiver can be derived. Repeating this procedure with the sample placed close to the vacuum window of the SIS receiver yields a new input noise temperature; the difference between the pair of measured input noise temperatures is directly proportional to the insertion loss due to the sample, which is accurate down to about 1%. Since this is a more sensitive method, we were able to measure thinner samples, the thicknesses of which are closer to what we will likely use for a vacuum window. Results utilizing this technique are comparable to those obtained with the QO-VNA for thicker samples. These results are included in Table 2.

TABLE 2
INSERTION LOSS AND LOSS TANGENTS OF SELECTED MATERIALS MEASURED WITH THE METHOD OF INTERSECTING LINES

Material	Thickness (mm)	Loss (%)	$\tan\partial \times 10^{-3}$
High Density Polyethylene (HDPE)	2.37	1.3 ± 0.2	0.9 ± 0.2
Honeywell Spectra (Dyneema)	1.84	4.7 ± 0.07	3.2 ± 0.1
Polypropylene (PP)	1.48	1.1 ± 0.3	0.8 ± 0.2
Z- Cut Crystalline Quartz	4.97	1.1 ± 0.03	0.5 ± 0.1

IV. LEAK RATE TESTS

We have also conducted vacuum leak rate tests of some selected material samples using a small chamber with a 51 mm diameter aperture. The vacuum window sample under test is pressed against the aperture O-ring, and a vacuum pump, connected to a shut valve, evacuates the chamber. Once a desirable base pressure is reached, the shut valve is closed and a pressure gauge monitors the rising internal pressure. Fig. 2 shows the results of the conducted leak-rate measurements, which is a relative measure of how effective the sample functions as a vacuum window.

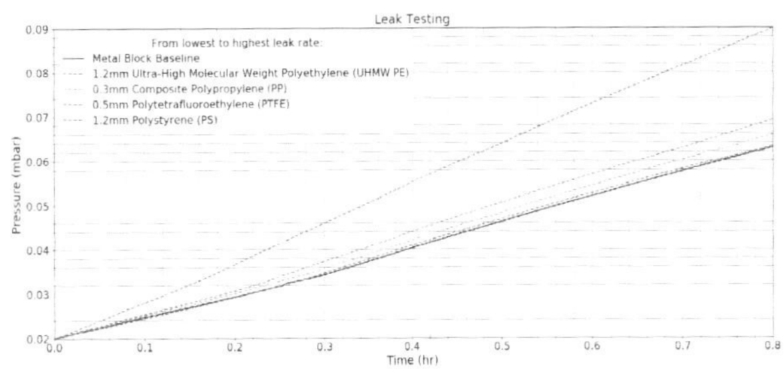

Fig. 2. Leak testing of various plastic materials on a 51mm diameter aperture vacuum chamber over a 0.8 hour time span.

V. CONCLUSION

Using the QO-VNA, we were able to determine the refractive indices and loss tangents of a variety of materials at around 300 GHz. The insertion losses of a number of sample windows have also been measured using the Method of Intersecting Lines with an SIS receiver. We have started to study the mechanical properties of these materials, including the leak rate and tensile modulus. Final selection of a window design will be a function of the refractive index, loss, mechanical strength, and the ease in which an effective anti-reflection coating can be implemented on the material.

REFERENCES

[1] P. Grimes, R. Blundell, S. Leiker, S. Paine, E. Tong, R. Wilson, and L. Zeng, "Receivers for the wideband submillimeter array," in *Proc. 31st IEEE Int. Symp. Space THz Tech.*, Tempe, AZ, March 2020.

[2] D. Barkats plus 77 other co-authors, "Ultra-thin large-aperture vacuum windows for millimeter wavelengths receivers", *in Proc. SPIE 10708,*

Millimeter, Submillimeter, and Far-Infrared Detectors and Instrumentation for Astronomy IX, 107082K, July 2018, doi: 10.1117/12.2312585.

[3] C. A. Balanis, *Advanced Engineering Electromagnetics*. Hoboken, NJ, USA: Wiley, 1989.

[4] R. Blundell, R.E. Miller and K. H. Gundlach, "Understanding noise in SIS receivers," Int. J. IR & MM Waves, vol. 13, pp. 3-14, Jan. 1992, doi: 10.1007/BF01011203.

[5] C.-Y. E. Tong, A. Hedden, and R. Blundell, "An empirical probe to the operation of SIS receivers — revisiting the technique of intersecting lines," *19th IEEE Int. Symp. on Space THz Tech.*, April 2008.

Contactless rotating MEMS waveguide switch for water detection at 557 GHz

Sofia Rahiminejad[1], Cecile Jung-Kubiak[1], Mina Rais-Zadeh[1,2], and Goutam Chattopadhyay[1]

Detecting water on other planets and heavenly bodies has been a long time goal of NASA. High-resolution heterodyne spectrometers are well suited to carry out such measurements. It can be used to detect unique molecular signatures, such as water molecules, with a high spectral resolution and precision over a wide range of wavelengths.

Calibration of the spectrometer is one of the critical aspect for such systems and typically a flip-mirror based calibration scheme is used to switch the signal between the antenna and the load for calibration of the receiver. Silicon-based radio frequency micro-electromechanical systems (RF MEMS) waveguide switches can be used for the same task, and take significantly less volume, mass and energy.

RF MEMS have shown to be compatible with waveguide components for submillimeter wave applications, since silicon micromachining allows for fabrication of 3D geometries, micrometer sized features, and high-aspect-ratio structures. In the past few years, waveguide components such as tunable capacitors, phase shifters, and waveguide switches have emerged for THz applications. Specifically, MEMS waveguide switches operating at frequencies from 400 GHz up to 750 GHz have been reported, with a 460 GHz single-pole double-throw waveguide switch demonstrated [1, 2].

To the best of the authors knowledge, all of the MEMS waveguide switches, developed to-date need electrical and mechanical contacts to block the wave, creating issues such as mechanical stress/stiction, poor life time, and if cycled many times, problems with ohmic contact resistance can arise. A contactless in-plane MEMS waveguide switch would therefore be greatly beneficial for THz applications.

In this paper, we present a contactless rotating MEMS waveguide switch operating between 500-750 GHz.

The switch has a U-bend waveguide attached to a rotating MEMS motor, as shown in Fig. 1a. The U-bend switches between connecting the Receiver (1) and the Antenna (2), or the Receiver (1) and the Load (3). By surrounding the U-bend waveguide with an electromagnetic bandgap (EBG) surface, the opposing waveguides do not need to be in contact with each other.

Simulations show that if there is a 10 μm air gap between the U-bend and the opposing waveguides, the isolation is larger than 30 dB, the insertion loss is less than 0.6 dB throughout the in-band, and the return loss at the receiver is larger than 25 dB between 500-722 GHz, and larger than 23 dB throughout the in-band.

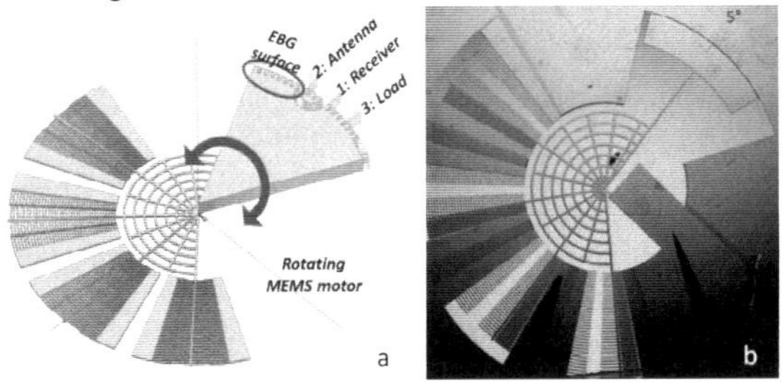

Fig. 1. a) A schematic of the MEMS motor with its four arms, defined in the 50 μm device layer of a SOI wafer, acting as levers to rotate the U-bend between two positions. The U-bend is surrounded by an EBG surface to confine the wave without any electrical contact to the opposing waveguides. b) The fabricated MEMS motor, shows a 5° rotation while actuated.

The MEMS motor has four arms that can be actuated to move together, both clockwise and counter clockwise. The MEMS motor was fabricated using an SOI (50 μm/2 μm/380 μm) wafer. The rotating motor is defined in the 50 μm device layer, while the U-bend waveguide is defined in the 380 μm handle layer. The U-bend waveguide needs to move 9° to switch between the two positions. The MEMS motor was measured to move up to ±5°, Fig. 1b.

The research described herein was carried out at the Jet Propulsion Laboratory, California Institute of Technology, Pasadena, California, USA, under contract with National Aeronautics and Space Administration.

References

[1] T. Reck, C. Jung-Kubiak, and G. Chattopadhyay. "A 700-GHz MEMS Waveguide Switch" IEEE Transactions on Terahertz Science and Technology, 6(4), 641–643, July 2016.

[2] T. Reck, C. Jung-Kubiak and G. Chattopadhyay, "A 460 GHz MEMS-Based Single-Pole Double-Throw Waveguide Switch," *2018 IEEE/MTT-S International Microwave Symposium - IMS*, Philadelphia, PA, 2018, pp. 773-775

[1] Jet Propulsion Laboratory, California Institute of Technology, Pasadena, CA 91109, USA.
[2] University of Michigan, Ann Arbor, MI 48105, USA.

Receivers for the wideband Submillimeter Array

Paul K. Grimes, Raymond Blundell, Patrick Leiker, Scott N. Paine, Edward C.-Y. Tong, Robert W. Wilson, and Lingzhen Zeng.

Abstract—**The Submillimeter Array (SMA) was conceived three decades ago as the world's first submillimeter interferometer capable of sub-arcsecond imaging in the frequency range from 200 to 700 GHz. Since it began full science operations in 2004 it has been continuously upgraded with new receiver cartridges, expanded intermediate frequency (IF) bandwidth, and augmented polarimetric and dual frequency observing modes. The next step in the development of the SMA is called the wideband Submillimeter Array (wSMA). The wSMA upgrade will replace the original SMA cryostats, receivers, and receiver selection optics with all new systems, and incorporate a number of major upgrades to the backend IF signal transport and correlator systems. This will enable the wSMA to operate with 32 GHz instantaneous bandwidth per receiver polarization, and will form the basis of future development efforts.**

Index Terms—**Radio astronomy, submillimeter astronomy, instrumentation, submillimeter receivers.**

I. INTRODUCTION

THE SMA began full science operations in 2004 with a suite of double side band (DSB) SIS mixer receivers operating in the 230, 345, and 690 GHz bands [2]. The 345 and 690 GHz receiver sets were co-polarized and could be used in conjunction with the 230 GHz receiver sets, which were operated in the orthogonal polarization. Each of the receivers had a 2 GHz-wide IF bandwidth and a purpose-built ASIC correlator combined the signals from the different antenna pairs for a total processed on-sky bandwidth of 2 polarizations x 2 sidebands x 2 GHz = 8 GHz.

Incremental improvements to the receivers and the addition of 240 GHz and 400 GHz band receivers that overlap with the original 230 GHz and 345 GHz bands on opposite polarizations, coupled with the development and deployment of a new correlator, have resulted in significant improvements in sensitivity, chiefly through increased on-sky bandwidth, and the addition of polarimetric imaging capabilities. The total processed on-sky bandwidth is currently 48 GHz, as the DSB receivers now provide output across a 12 GHz wide IF band (4–16 GHz) and the sidebands are separated in the correlator. The additional correlator capacity to process this bandwidth is currently being commissioned.

TABLE I
SMA VS WSMA RECEIVER SPECIFICATIONS

	SMA	wSMA
Receiver Bands (LO Tuning, factors from phase lock loop frequency)	**200** 186-242 GHz, ×2 multiplication **240**: 210-270 GHz, ×3 multiplication **300**: 271-349 GHz, ×3 multiplication **400**: 330-420 GHz, ×4 multiplication	**230**: 210-270 GHz, ×3 multiplication **345**: 280-360 GHz, ×3 multiplication
Polarization	Two single polarized receivers from two groups (**200**+**300**, **240**+**400**) 45° to elevation axis.	Dual polarized receivers 45° to elevation axis.
Receiver Selections	**200** + **240** **200** + **400** **300** + **240** **300** + **400**	**230** dual polarized **345** dual polarized **230** pol 1 + **345** pol 2 (with wire grid) **230** dual pol + **345** dual pol (with dichroic, limited tunings)
IF Band	4-16 GHz	0.1-4 GHz + 4-16 GHz (possible expansion to 20 GHz)

Comparison of the current receiver configuration of the SMA (as of May 2020) and the proposed wSMA receiver configuration.

The receivers offer good performance over a wider range of sky frequencies than the local oscillator (LO tuning ranges shown in this table. However, performance at the extreme edges of the accessible sky frequencies is not guaranteed.

Following these improvements, the next step in the ongoing development of the SMA will be to replace the original receiver cryostats and receiver selection optics with all-new hardware [3]. A major impetus for this upgrade is the need to replace the aging cryogenic systems, which are no longer supported by the

The Submillimeter Array is a joint project between the Smithsonian Astrophysical Observatory and the Academia Sinica Institute of Astronomy and Astrophysics and is funded by the Smithsonian Institution and the Academia Sinica.

Paul K. Grimes, Raymond Blundell, Patrick Leiker, Scott N. Paine, Edward C.-Y. Tong, Robert W. Wilson, and Lingzhen Zeng are with the Center for Astrophysics | Harvard & Smithsonian, Cambridge, MA 02138, USA (e-mail: pgrimes@cfa.harvard.edu).

Fig. 1. SMA receiver and receiver optics cage in an SMA antenna, circa 2005. (with John Barrett and SMA Director Raymond Blundell). The SMA beam waveguide mirrors M6 and M5 are visible at the top of the image, with the optics cage containing the moving wire grid and mirror combiners below the cross-beam. The cryostat is below the optics cage, with the receiver electronics "cheeks" mounted around it.

manufacturer. This necessary replacement provides an opportunity to redesign the receiver systems to take advantage of recent developments in submillimeter receiver design, to make provision for future receiver developments, and to optimize the receiver systems for the key science goals of the SMA for the coming decade [4].

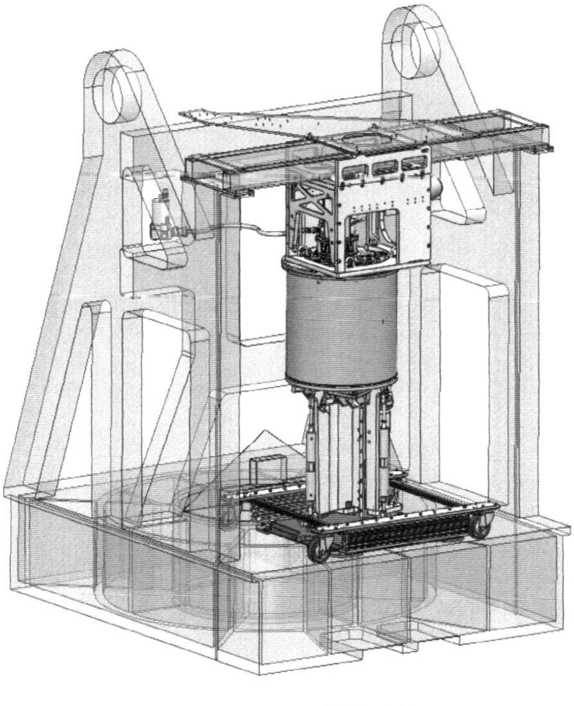

Fig. 2. wSMA Receiver cryostat (vacuum shell in pink), support structure and alignment structure in the SMA receiver cabin (existing structures are transparent). Rendering courtesy of HPD Inc.

A side-by-side comparison of the main receiver specifications for the current SMA and the wSMA is given in Table 1.

The increased receiver IF bandwidth, improved throughput due to the reduced use of room temperature and transmissive optics, incremental improvements to our SIS mixer performance, and improved coalignment of the orthogonal polarization beams on the sky introduced by this upgrade are expected to result in a significant improvement in the sensitivity of the SMA, as well as improved instrumental polarization and efficiency for polarimetric observations, and improved reliability of the receiver systems.

The upgrades to the wSMA receiver and other systems will also serve as a basis for future development of the SMA. The wSMA receiver system has been designed to allow the future development and deployment of sideband-separating receivers, and upgrades to even wider IF bandwidths, while the upgrades to the IF/LO signal transport system will provide communication capacity for digital IF transmission and other advancements.

One particular feature of the wSMA cryostat and optics design is the availability of space and an optical port to allow the installation of additional "guest" receivers alongside the wSMA receivers. It is envisioned that these guest receiver positions could be used to demonstrate new receiver technologies in an interferometer environment (e.g. multi-beam receivers or receivers using new frontend technologies), or to deploy specialized receivers for dedicated observing campaigns.

In this paper, we give an update on the status of the wSMA upgrade, describe the wSMA receiver systems, and present the key receiver technology developments that are part of the upgrade. We also present current developments for improving the calibration of the SMA/wSMA, incorporating a sky frequency dependent system noise temperature measure.

II. The wSMA Receiver Upgrade

The replacement of the SMA receivers, receiver optics, and cryogenics with a completely new receiver system is the major element of the wSMA upgrade. The existing receiver system is shown in Fig. 1, and consists of the receiver optics cage, mounted on top of the receiver cryostat, with the whole stack supported from the floor of the antenna cabin. The receiver optics cage mates to the "M456" plate that carries the SMA beam waveguide, made up of M4, M5 and M6, and the calibration load unit. This plate is supported in the receiver cabin by a beam across the cabin.

The new wSMA receiver system will replace everything between the M456 plate and the cabin floor (Fig. 2). The new cryostat will be aligned to the M456 plate by an alignment structure on top of the cryostat, and will be mechanically supported from the cabin floor by a sprung support that will ensure that the alignment structure is in compression at all times.

The new wSMA receiver cryostats will each house two dual polarization receiver cartridges, operating in two frequency bands; a low band with a local oscillator (LO) frequency from

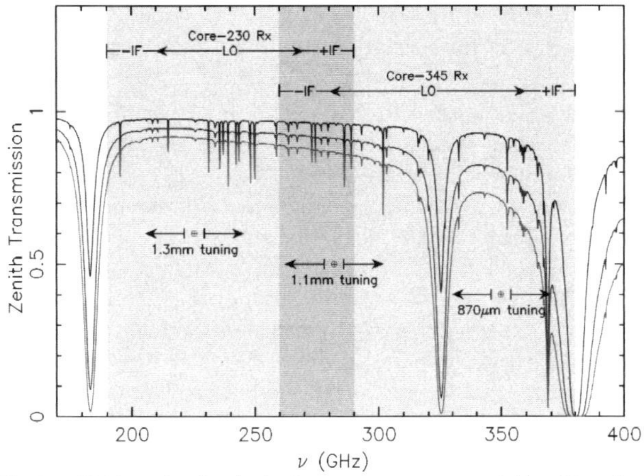

Fig. 3. wSMA receiver bands sky frequency coverage, overlaid on Maunakea sky transmission for the best 50%, 25% and 10% weather.

210 to 270 GHz, and a high band with an LO frequency from 280 to 360 GHz. The receiver system will operate with an IF of 0.1 – 16 GHz (potentially expanding to 0.1–20 GHz at a later time), giving additional frequency coverage outside of the LO tuning ranges so that continuous coverage will be obtained from ~194 to ~374 GHz. The tuning range of the wSMA receivers is shown in Fig. 3. The receivers will deliver 16 GHz bandwidth per polarization per sideband, for a total of 2x2x16 = 64 GHz of processed on-sky bandwidth.

The current room temperature receiver selection optics will be replaced with a receiver selector wheel internal to the cryostat and mounted on the 50 K radiation shield. This will allow the incoming beam to be directed to either individual receiver cartridge for dual polarization single band observing, or to split the orthogonal polarizations of the incoming beam between the receiver cartridges for dual band observing. A fourth position of the selector wheel will be available for a future cold dichroic beamsplitter to allow dual-band, dual-polarization observing.

Each wSMA receiver cartridge will use a single dual-polarized feedhorn and an orthomode transducer feeding two DSB SIS mixers to receive the two polarizations. The local oscillator signal will use waveguide LO injection in the 4 K receiver front-end module.

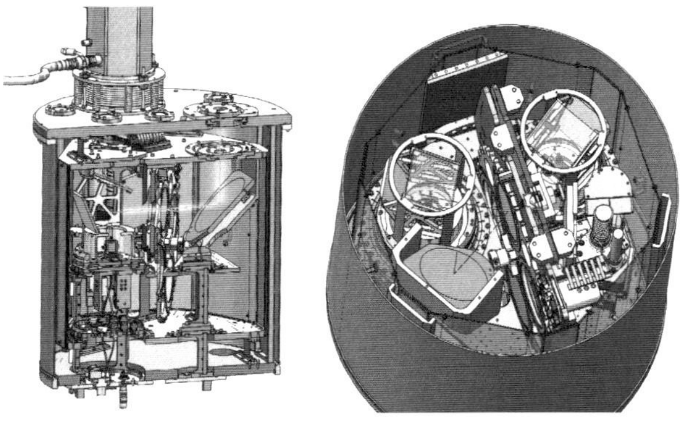

Fig. 4. Cut-away CAD renderings of the wSMA cryostat, showing the cryostat structure and the optical layout (optical path highlighted in blue on the left image). Renderings courtesy of HPD Inc.

Fig. 5. Photographs of one of the two prototype cryostats under construction at HPD Inc. (left) Complete cryostat on the support structure, with the cryostat vacuum jacket lowered and radiation shields removed to allow access to the cryostat internals. The alignment structure that positions the cryostat with respect to the SMA beam waveguide is in place on top of the cryostat, with the optical alignment telescope used for checking the alignment of the cryostat optics at the very top of the image. (center) Zoom in on the cryostat internals. The pulse tube cooler can be seen on the left of the image, with one of the two receiver cartridges front and center. (right) Receiver cartridge, with the return mirror used to test optical alignment in place. Photographs courtesy of HPD Inc.

A. Receiver cryostat

The wSMA receiver cryostat (Fig. 4) will hold the two receiver cartridges, the cold receiver selection optics and a cold calibration load. The cryostat will be cooled by a Cryomech PT-407RM or PT-410RM pulse-tube cooler driven by a variable frequency compressor unit. In contrast to the Gifford-McMahon backed Joule-Thomson coolers used in the current SMA cryostats, the remote motor pulse-tube cooler has no moving parts within the cryostat, and so should be significantly simpler to maintain.

The two receiver cartridges will be inserted from the base of the cryostat, with all receiver wiring, IF signal cables, and LO waveguides passing along the receiver cartridge to the receiver front-end. Thermal connections to the receiver cartridges will be via automatic thermal links. Installing a receiver cartridge should not require any connections to be made inside the cryostat.

The detailed cryostat design and fabrication of two prototype cryostats is being carried out by HPD Inc. (Boulder, CO). The design of the prototype cryostats has been completed, and the two prototype cryostats are now undergoing testing at HPD Inc. before delivery later this summer. Fig. 5 shows photographs of the prototype cryostat and receiver cartridge.

After acceptance tests, one prototype cryostat will be used to integrate and test wSMA receiver prototypes in the SMA Receiver Lab in Cambridge, MA, while the other will be shipped to the SMA site on Maunakea for fit and compatibility testing in an SMA antenna.

B. Receiver optical design

The quasioptical beam from the telescope is fed to both receiver cartridges via a single vacuum window and IR filter. Inside the cryostat, the beam is folded through 90° before

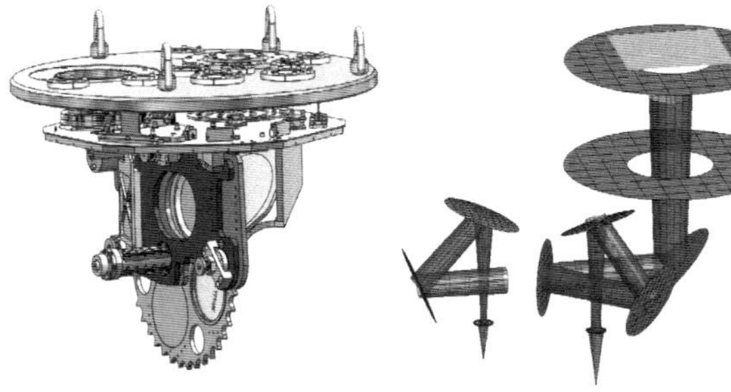

Fig. 6. (left) CAD rendering of the cryostat fold mirror and selector wheel supported from the upper vacuum plate and upper radiation shield plate of the cryostat. Rendering courtesy of HPD Inc. (right) GRASP model of the cryostat and receiver optics showing the Gaussian beam envelopes as the beams from each receiver feed pass through the receiver optics, selector wheel, fold mirror, and IR filter and vacuum window apertures.

reaching the receiver selector wheel. The selector wheel is a four-position wheel driven by a cryogenic stepper motor which can position one of four optical elements in the incoming beam. By choosing between an open port, a plane mirror and a wire grid, one or both linear polarizations of the incoming beam can be fed to either receiver cartridge. See Fig. 4, (left) and Fig. 6, (right) for illustrations of the optical layout.

The fourth position of the selector wheel will be used at a later time for a cold dichroic beamsplitter, to allow both polarizations to be fed to the two receiver channels simultaneously. Due to the close spacing in frequency of the two receiver bands, the receiver tunings available with the dichroic splitter will likely be limited to the lower end of the Low receiver and the upper end of the High receiver bands. This would allow, for example, the CO(1–2) and CO(2–3) lines at 230 GHz and 345 GHz to observed simultaneously in both polarizations, but will not allow the full frequency coverage of each receiver to be used.

From the selector wheel, the beam passes to the receiver optics on each receiver cartridge, consisting of a clamshell mirror pair that focuses the incoming beam on to the feed horns of the receiver front ends. Details of the cryostat and receiver optical design are given in [5].

The fold mirror and selector wheel optics inside the cryostat are mounted on the top plate of the cryostat radiation shield, and cooled to ~60 K. The receiver optics and receiver front end are mounted to the top of the receiver cartridges on sprung "floating 4 K plates" which are thermally strapped to the receiver cartridge cold plate that is in turn thermally linked via the automatic thermal link to the cryostat second stage cold plate. The floating part of the receiver cartridge is aligned on receiver insertion by a G10 structure that mates via a kinematic interface to the radiation shield top plate. The cryostat radiation shield top plate thus effectively serves as an optical bench that controls the alignment between all of the cooled optics and feeds.

C. Receiver cartridge and front-end design

Each wSMA receiver cartridge (Fig. 7, left) carries the receiver optics and a single front-end receiver module on the floating 4K plate, and up to four IF LNA/isolator amplifier modules on the underside of the fixed 4K plate.

The front-end receiver module (Fig. 7, right) consists of a single dual-polarized feedhorn and an orthomode transducer (OMT) feeding two DSB SIS mixers to receive the two polarizations. The local oscillator signals are coupled in to the signals from the OMT outputs through hybrid waveguide/planar circuit directional couplers and the combined signals are then fed to double-sideband SIS mixers.

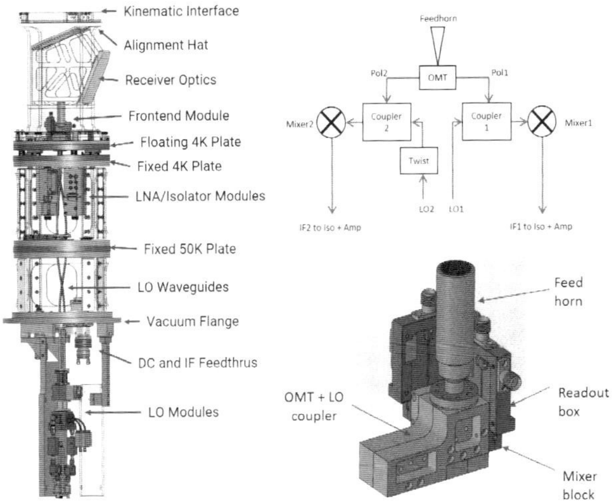

Fig. 7. (left) CAD rendering of populated wSMA receiver cartridge, with LO module installed. (upper right) Schematic of the wSMA receiver frontend, and (lower right) CAD rendering of a wSMA receiver frontend module.

The SIS mixers are an evolution of the wide IF band SIS mixer designs currently in use on the SMA. The design of these mixers is discussed in [6].

For the wSMA, we are proposing to incorporate a split IF output scheme to allow an IF output band from 0.1 GHz to 16 GHz or even higher. This scheme was discussed in [7], and uses a diplexer as the first element in the IF circuit, which splits the IF signal from the SIS mixer into two channels. The main 4–16 GHz IF channel will use a similar cryogenic isolator and Low Noise Factory InP HEMT LNA to that used on the SMA. The 0.1–4 GHz IF channel will be amplified by a cryogenic SiGe LNA. After amplification, the two IF channels will be brought out of the cryostat and independently transported and processed.

The front-end module (Fig. 7, lower right) will be constructed from a few integrated blocks. The profiled corrugated feed will be electroformed, and mounted on to the OMT-coupler module. The OMT and LO waveguide coupler will be built as an integrated module, with two mixer modules mounted on the outputs. The IF diplexer and bias tee for the SIS mixer will be built into the "readout box" module mounted on sides of the two mixer modules.

Further details of the front-end receiver design and the design and prototyping of the various components of the front-end are given in [8].

Local oscillator signals are injected into the signal path by waveguide directional couplers. The LO signals are fed to the front-end from the Receiver Cartridge vacuum flange using overmoded WR-10 stainless steel waveguides. The local oscillator units will be mounted to the outside of the Receiver Cartridge vacuum flange.

Both receiver bands will use LO sources based on amplifier-multiplier chains (AMC) driven by voltage-controlled oscillator (VCO) fundamental sources. After multiplication and amplification of the fundamental signal to W-band, a small portion of the generated signal is coupled out to a harmonic mixer and phase-lock loop system, which locks the LO signal to the master reference generator signals generated in the SMA control building and transmitted to all the antennas. The final multiplication to the receiver LO frequencies is then carried out by a x3 multiplier, after which the signal level is controlled by an optically controlled silicon chip attenuator [9].

Achieving IF operation down to 0.1 GHz will require significant care in the design of the AMCs in order to control excess LO sideband noise and ensure that it does not adversely affect receiver noise temperatures at low IF's. Some of the design procedures for our AMC LOs are discussed in [10].

Initially the two receiver bands will use somewhat different LO schemes. The 230 band receiver will use two independent LO AMC chains, one for each polarization, while the High band receiver will use a single LO AMC chain, with a power splitter and independent attenuators just outside the cryostat to drive the two polarization channels.

The ability to independently tune the two polarizations of the Low band receiver will be particularly useful for spectral line surveys, allowing near continuous frequency coverage over 64 GHz of bandwidth in the 194–286 GHz frequency range accessible by the Low band receiver. The High band frequency band is divided approximately in half by the deep atmospheric absorption line at 325 GHz. This line makes split tuning of the two polarization channels less useful for this band. With IF operation from 0.1–16 GHz, almost the entire useful atmospheric window between the 325 GHz and 380 GHz water vapor lines can be accessed with a single LO tuning. The High band LO units can easily be upgraded to dual AMCs with independent operation at a later date.

III. Upgrades To Other Systems

Fully utilizing the capabilities offered by the upgraded wSMA receivers also requires upgrading a number of other systems, particularly in the IF signal transport and correlator for the array.

1) IF/LO Signal Transport

The wider 16 GHz IF bandwidth being deployed now, and specified for the wSMA receivers, requires that the current analog RF-over-fiber based IF signal transport system be certified to handle that bandwidth.

A number of microwave components in the IF signal processing systems in the both the antennas and control building's correlator room have been upgraded to 18 GHz or higher components. The Ortel RF-over-fiber receiver/transmitter pairs operating at 1310 nm have been checked for performance to 16 GHz, and the few units performing below requirements were replaced with spares.

The entire IF signal transport system is now ready for 4–16 GHz analog RF-over-fiber operation. This capability is being used for the current commissioning of the 12–16 GHz IF expansion.

The proposed 0.1–4 GHz low IF expansion will also be transmitted over the same analog fiber links, either recombined into the existing channel; or transmitted on an adjacent CWDM channel, with a separate IF processor in the antenna cabin.

Currently the SMA transmits the Master Reference Generator signals for each of the two active receiver's LO references by multiplexing them onto the same analog RF-over-fiber link. This means that care must be taken when choosing tunings to avoid interference between the two MRG signals. As part of the reconfiguration of the IF for the wSMA upgrade, these two signals will be separated onto different WDM optical channels, removing the possibility of interference between them.

2) Correlator Expansion

Expanding the SMA's IF bandwidth to 4–16 GHz necessitates building more correlator capacity to handle this additional bandwidth. The expansion of the SWARM correlator [11] to handle the additional 12–16 GHz IF band is mostly complete, with two additional SWARM segments (originally called quadrants, but now six in number) currently being commissioned, each handling 2 GHz of bandwidth from both receiver channels. Additional block down-convertors have been purchased and installed, and the final steps in software deployment and science verification of the expanded correlator are now being undertaken.

Further expansion to handle the 0.1–4 GHz IF band is being planned. We expect to test the split IF system over a limited bandwidth on several SMA receivers during the latter part of this year, with an additional block downconverter, using the spare SWARM ROACH components that have recently been purchased. The full deployment of the 0.1–4 GHz IF band will occur alongside the wSMA receiver deployment.

3) Proposed next generation correlator development

With the deployment of wSMA receivers capable of wider IF bandwidths, from 0.1 to at least 20 GHz IF, and simultaneous operation of four receiver channels (dual receiver, dual polarization) at the same time using a dichroic splitter, a new correlator and IF signal transport concept is required to make use of the full capabilities of the receivers.

Concepts for a "next generation SWARM" correlator are currently being developed. Such a system will likely digitize the IF signals in the antenna cabins before transmitting data packets over high speed Ethernet links running over multiple DWDM optical fiber channels. Such a system would take advantage of the recent developments in extremely high speed Ethernet links for the general telecoms industry, and the signal transport would be based on commercial off-the-shelf technology.

IF digitization in the cabin and transmission via packetized networks on the SMA would be an enabling development for future receiver upgrades, including deployment of wide IF band sideband separating receivers, and multi-beam and other guest receivers, both of which require significantly increased flexibility in the IF signal transport and processing.

IV. Wideband Calibration

With a wide instantaneous bandwidth, as high as ~20%, the system gain and sensitivity can vary significantly as a function of observing frequency. In addition, for a double-sideband system, the sideband ratio can vary with IF frequency.

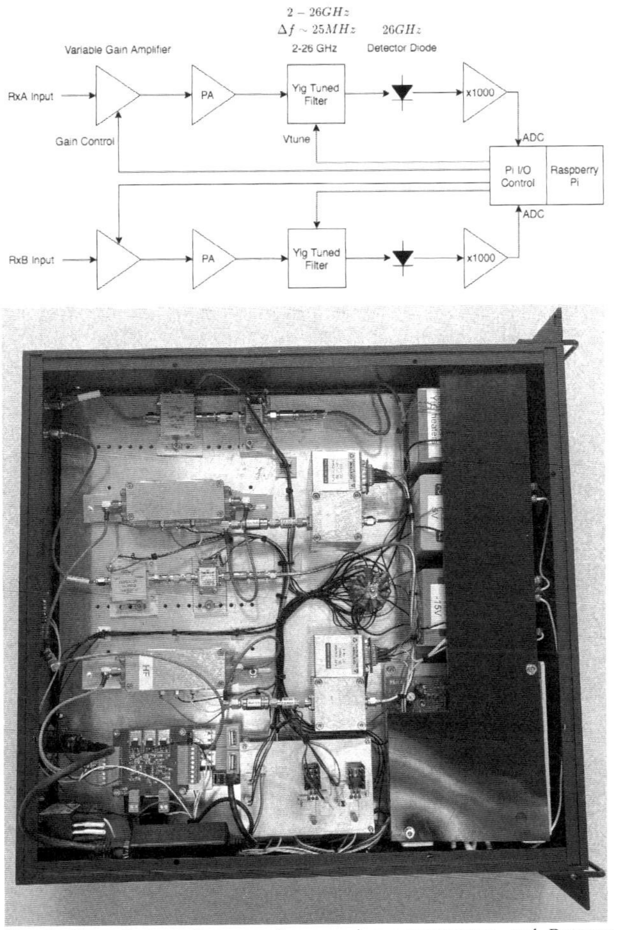

Fig. 8. Top: System diagram of a scanning spectrometer, and Bottom: photograph of the internal construction of a scanning spectrometer. RxA and RxB inputs (top left in photo) are connected to the two IF signal paths of the SMA, with one of the two operating receivers feeding each signal chain. The Raspberry Pi computer (under interface board on bottom left) controls the gain of the first amplifier and the tuning of the YIG filter, records the detector diode output, and outputs it to the SMA's control system.

Effective use of a wide-band heterodyne instrument requires knowing the system gain and sensitivity as a function of sky frequency, so that appropriate weightings can be assigned to frequency channels, and for accurate determination of spectral line ratios for different lines that appear within the bandwidth of the receivers.

For the current expansion of the SMA to 4–16 GHz IF observations, we have developed and deployed "scanning spectrometer" instruments to each of the SMA antennas. Each scanning spectrometer can determine the DSB system temperature (T_{sys}) and relative gain of the two active receiver channels as a function of IF frequency by recording spectrally resolved IF power measurements on the SMA's ambient calibration load and on the sky.

The system diagram of a scanning spectrometer is shown in Fig. 8. Each scanning spectrometer signal chain consists of a variable gain amplifier and power amplifier for gain control, a Yttrium iron garnet (YIG) tuned filter with 25 MHz bandwidth that can be swept from 2–26 GHz, a detector diode and DC amplifier, and a Raspberry Pi controlled ADC/DAC unit. By

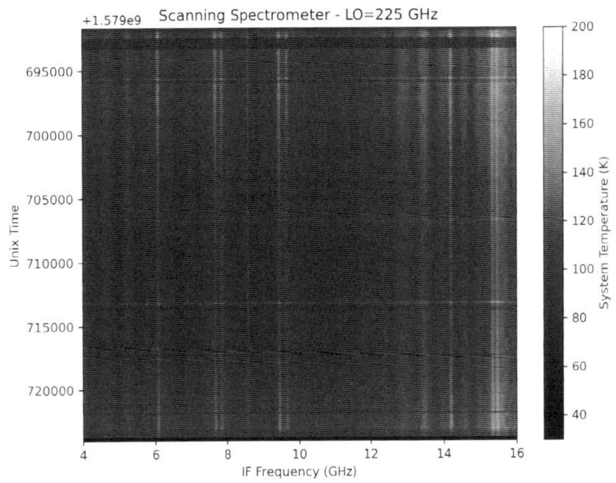

Fig. 9. Scanning spectrometer DSB Tsys data for one frequency recorded during a test observation. Vertical lines in the data correspond to stratospheric ozone lines in one of the two sidebands. Sharp horizontal lines correspond to rapid changes in the telescope pointing to switch to flux calibration and pointing targets, while slow variations track changes in atmospheric conditions and observation target elevation.

sweeping the YIG filter, a DSB IF spectrum can be recorded across the current 4–16 GHz IF bandwidth of the SMA in around one second.

In normal observations, an IF spectrum is taken on the ambient load every time it is inserted for routine calibrations. While observing the (colder than ambient) sky, IF spectra are taken continuously at a rate at least as fast as the SMA's correlator integration rate. The T_{sys} of each receiver as a function of IF can thus be determined for each of the integrations during an observation, using the usual Y-factor method. An example of the scanning spectrometer data for one observation and receiver is shown in Fig. 9.

Since the SMA correlator separates the visibilities from the DSB receivers into two sidebands via Walsh switching, ideally we would like to determine a sideband-separated T_{sys} to correctly calibrate the sideband-separated visibilities. The broad bandwidth of the SMA typically encompasses a number of strong ozone lines, which in principle provide enough predictable spectral structure to retrieve a low-resolution sideband-separated spectral T_{sys} from the scanning spectrometer data.

The scanning spectrometer system is currently undergoing testing and integration into the SMA data acquisition system. It is expected that data from the scanning spectrometer system will be available when the SWARM correlator's 12–16 GHz extension comes online for routine observations later this year.

V. wSMA Deployment Timeline

Development and deployment of the wSMA upgrade will take several years. Over the next year, we expect to receive the two prototype cryostats, test fit one to an SMA antenna, and integrate prototype front-end receivers in the prototype cryostats, before placing an order for the full complement of production cryostats. At the same time, numerous developments will be tested and deployed on the SMA, including the Raspberry Pi based controllers for receiver optics and local oscillator units, software upgrades, and IF/LO signal transport upgrades.

Some initial elements of this upgrade program have already been completed, and are due to be released for production observing later this year. In particular, the expansion of the SMA IF bandwidth to cover an additional 4 GHz of bandwidth in the 12–16 GHz range is currently undergoing final science verification before being offered to observers in the next semester on a shared-risk basis. The scanning spectrometer systems are mostly deployed, and should be available at the same time.

Following the testing of the prototype receivers and cryostats, we expect that it will take 2–3 years to receive, integrate and deploy the wSMA receiver systems to all SMA antennas. To upgrade each antenna, the antenna must be taken into the SMA hangar to remove the existing receiver and other ancillary systems, to carry out scheduled maintenance of the antenna, and to fit the new receivers and support systems. After this, significant testing will be needed to verify the performance and calibration of the new receivers before the antenna can be incorporated into the SMA for regular array observations.

Throughout this transition period, compatibility between the upgraded and unmodified antennas will need to be maintained. To do this, we will limit normal science observations to the range of frequencies (sky and IF) that are common to both sets of antennas. The polarization basis for the new receivers has been chosen to match that of the existing receiver cartridges, and careful selection of receiver combinations and MRG frequencies will be required to successfully operate the two receiver types together.

Following the upgrade of all (or most) antennas and science verification, the extra capabilities of the wSMA will be available to all SMA observers.

References

[1] Ho, Paul T. P., James M. Moran, and Kwok Yung Lo, "The Submillimeter Array.", The Astrophysical Journal, 616 (1), 2004

[2] Raymond Blundell, "The Submillimeter Array – Antennas and Receivers", *15th Int. Symp. Space THz Tech.*, Northampton, MA, 2005.

[3] Paul K. Grimes, Raymond Blundell, Scott Paine, John Test, C-Y. Edward Tong, Robert W. Wilson, and Lingzhen Zeng, "The wSMA receivers – a new wideband receiver system for the Submillimeter Array", *28th Int. Symp. Space THz Tech.*, Cologne, Germany, 2017.

[4] ed. D. Wilner contributing authors: E. Keto, G. Bower, T.C. Ching, M. Gurwell, N. Hirano, G. Keating, S.P. Lai, N. Patel, G. Petitpas, C. Qi, TK Sridharan, Y. Urata, K. Young, Q. Zhang, J.-H. Zhao, "Science with the wideband Submillimeter Array: A Strategy for the Decade 2017–2027", *SMA Memo Series*, 165, 2017

[5] Paul K. Grimes, Scott N. Paine, Lingzhen Zeng, and Edward C.-Y. Tong, "Optics and Feed Design for the wSMA Receiver System", *31st Int. Symp. Space THz Tech.*, Tempe, Arizona, 2020.

[6] C. E. Tong, P. Grimes, R. Blundell, M. Wang and T. Noguchi, "Wideband SIS Receivers Using Series Distributed SIS Junction Array," in IEEE Transactions on Terahertz Science and Technology, vol. 3, no. 4, pp. 428-432, July 2013, doi: 10.1109/TTHZ.2013.2259624.

[7] Raymond Blundell, Robert Kimberk, Edward C.-Y. Tong, Paul K. Grimes, Nathan Hagos, Lingzhen Zeng, "A 1.3 mm Superconductor Insulator Superconductor Mixer Receiver with 40 GHz Wide Instantaneous Bandwidth", *30th Int. Symp. Space THz. Tech.*, Gothenburg, Sweden, 2019.

[8] Lingzhen Zeng, C Edward Tong and Paul K. Grimes, "Waveguide Components for wSMA Frontends", *31st Int. Symp. Space THz. Tech.*, Tempe, Arizona, 2020.

[9] J. A. Connors, C.-Y. E. Tong, P. K. Grimes, A. Gebreyohannes and L. Liu, "Design and Measurement of a Waveguide Probe Based WR3.4 Optically Controlled Modulator", *28th Int. Symp. Space THz Tech.*, Cologne, Germany, 2017.

[10] C.-Y. E. Tong, P. K. Grimes, C.-C. Han, "Design Considerations for Amplifier-Multiplier Chain (AMC) for Low Noise Local Oscillator", *27th Int. Symp. Space THz Tech.*, Nanjing, China, 2016.

[11] Rurik A. Primiani, Kenneth H. Young, André Young, Nimesh Patel, Robert W. Wilson, Laura Vertatschitsch, Billie B. Chitwood, Ranjani Srinivasan, David MacMahon and Jonathan Weintroub, "SWARM: A 32 GHz Correlator and VLBI Beamformer for the Submillimeter Array", *Journal of Astronomical Instrumentation*, vol 5, 4, 2016.

A Horn-coupled 4-beam Dual-polarization Balanced SIS Mixer Based on Planar-integrated Circuits

Wenlei Shan[1], Shohei Ezaki[1], Keiko Kaneko[1], Akihira Miyachi[1], Takafumi Kojima[1], and Yoshinori Uzawa[1]

Coherent focal plane array receivers are employed in radio astronomical observations for imaging celestial objects with extended structures, whose size is much larger than the angular size of the telescope beam. However, in mm/sub-mm regime the complexity of coherent receiver frontends, which are conventionally constructed with metal waveguide circuits, imposes a limit on the number of pixels arrayed in the focal plane of a radio telescope and results in a narrow field of view. We have been developing a conceptual solution to enable compact focal plane heterodyne detector arrays with SIS mixers for wide field-of-view astronomical observation at mm and sub-mm wavelengths. The key ideas of this concept include (1) the exploitation of planar integrated circuits (ICs) to allow the use of compact planar orthomode transducers and hybrid bridge used in a balanced configuration (either a balanced mixer or sideband separation mixer); (2) the adoption of a semi-two-dimensional buried-in LO distribution network; and (3) the implementation of LO coupling between the LO distribution network and the mixer ICs by using membrane-borne waveguide probes. We have demonstrated the feasibility of THz superconducting integrated circuits with a single-pixel prototype receiver [1-2].

After that, we have been working on the development of a compact 4-beam horn-coupled dual-polarization balanced SIS mixer based on aforementioned ICs operating at 2 mm wavelengths. This experiment complements the conceptual study by demonstrating a fully functioning array with implementing all the key ideas in the concept. A photo of the receiver frontend is shown in Fig. 1. The lateral size of the frontend is limited by the horns and can be further smaller if the aperture size of the horn antenna reduces. This compactness is not achievable with the conventional integration method.

This presentation reports the uniformity of the LO distribution, followed by the RF performance of the array receiver. Moreover, a comparison is made between the performance of a mixer in the array context and that in a single-pixel mixer block. With this prototype receiver, the experimental methods and issues that are unique to a densely clustered SIS multibeam receiver are investigated.

[1] National Astronomical Observatory of Japan, Tokyo, 181-8588, Japan.

Fig. 1. Photo of the 4-beam mixer holder mounted inside the testing cryostat. Signals of 2 mm wavelengths are coupled through four corrugated horns attached at the upper side. The single LO injection port is seen at the near-side of the block. IF/DC are conducted by using 16 coaxial cables.

References

[1] W. Shan, S. Ezaki, J. Liu, S. Asayama, and T. Noguchi, "A New Concept for Quasi-Planar Integration of Superconductor-Insulator-Superconductor Array Receiver Front Ends," *IEEE Trans. on Terahertz Sci. and Tech.*, vol. 8, no. 4, 472-474, 2018.

[2] W. Shan, S. Ezaki, K. Kaneko, A. Miyachi, T. Kojima and Y. Uzawa, "Experimental Study of a Planar-Integrated Dual Polarization Balanced SIS Mixer," *IEEE Trans. on Terahertz Sci. and Tech.*,, vol. 9, no. 6, pp. 549-556, Nov. 2019.

Advanced Tuning Algorithms for High-Frequency SIS Mixers

R. Hesper[1], J. Barkhof[1], A.M. Baryshev[1], T. Vos[1], G. Siringo[2], N. Phillips[3] and P. Yagoubov[3]

Three tuning parameters are key to SIS mixer performance: SIS bias voltage, LO pumping level and Josephson current suppression. The former two are normally determined by an iterative optimization process, although there are a couple of pitfalls. The suppression of the Josephson current, however, is a much more complicated matter. This is especially the case in the high-current-density AlN-barrier junctions that populate most of the ALMA Band 9 receivers (600-720 GHz), as these tend to show hysteretic or multiple-regime behavior for the Josephson current suppression as a function of external applied magnetic field. The Band 9 mixers have other issues, intrinsic to the device physics, as well: since the gap frequency of niobium (~680 GHz) lies within the band, tuning structures get progressively lossy, with a strong dependence on the magnetic field around the point where the superconducting order parameter collapses. Also, the Band 9 junctions use a Al-AlN-Nb trilayer on top of Nb, which means that the superconductivity of the bottom Al layer (with a native Tc of around 1K) is purely induced by the proximity effect, also introducing a strong magnetic field effect. Together, these effects indicate that one should strive for the lowest possible magnetic Josephson suppression field in order to obtain the highest noise performance, especially in the upper part of the band.

The tuning parameters that were supplied with the 186 ALMA Band 9 mixers were biased towards safer and more repeatable regimes rather than ultimate performance, for operational reasons. Here we present the results of the recently concluded ALMA Band 9 Advanced Tuning Study commissioned by ESO. The objectives of this study were to develop "intelligent" automatic tuning algorithm to enable receivers to operate in more critical regimes that were previously avoided, and to investigate the corresponding improvement in noise performance. Secondary benefits of the automatic tuning algorithm are to reduce ALMA array operator's time to find and confirm the tuning parameters for the entire array and to minimize the number of mixers that would be flagged as unusable due to tuning problems during observations.

In the framework of this study, we first had to develop a new software infrastructure. Our original engineering software, while excellently capable of qualifying production receivers (as proven by the successful ALMA Band 9 and Band 5 production campaigns), is not very suitable for adaptive "intelligent" algorithms because of the absence of conditional statements and loop constructions. The new system we developed is structured as a series of extension libraries of the Python language. This gives the full power of an established high-level programming language suitable for implementation of any adaptive or interactive algorithm one can think of.

Using this infrastructure, we first reproduced and fully automated the formerly semi-automatic and "eyeball" algorithms, arriving reliably and repeatably at very similar tuning points as with the classical methods. Working from there, we investigated incrementally several improved and even completely different tuning methods. The final implementation is a three-stage algorithm for optimizing the Josephson suppression: an initial approximation by finding the first minimum in the critical current around zero-bias, followed by a refinement based on the integrated IF power of the Josephson power, and finally a qualification by the width of the Josephson feature. This procedure is preceded by a demagnetization-defluxing step and together they form a loop that is repeated until a qualified Josephson current suppression is found.

As stated before, the delivered Band 9 mixers were supplied with "safe" tuning parameters, meaning in practice that the Josephson suppression current is set to the second minimum. The new algorithm, however, can tune most of the mixers reliably in the first minimum. Together with the almost linear relationship that we find between the achievable noise temperatures and the applied magnetic field, this leads to a significant sensitivity improvement in the set of re-tested mixers, compared to the traditional second-order suppression. We checked, of course, whether the mixers still conform to other important performance specifications (notably the amplitude stability and small-signal gain compression level).

By applying a simplified non-interactive tuning algorithm to the archived qualification data of the delivered mixers, we can make a good estimate of the expected performance improvement when the ALMA Band 9 array is tuned with the new algorithm: in about 80% of the mixers the noise could be improved by 15-30K (from an average of about 100K) in the top part of the ALMA Band 9 frequency range (which contains the CO J=6-5 line at 691 GHz), 10-25K in the middle and 4-7K at the bottom of the band.

[1]NOVA Sub-mm Instrumentation Group, University of Groningen, Landleven 12, 9747 AD Groningen, The Netherlands

[2]Joint ALMA Observatory, Vitacura, Santiago de Chile, Chile

[3]European Southern Observatory, 85748 Garching bei München, Germany

NOTES:

Optics and Feed Design for the wSMA Receiver System

Paul K. Grimes, Scott N. Paine, Lingzhen Zeng, and Edward C.-Y. Tong.

***Abstract*— We are currently developing a major upgrade for the Submillimeter Array (SMA) receiver systems, which will involve the complete replacement of receiver cryostat, cartridges, front-ends, and receiver optics with a newly designed receiver system. This upgrade, known as the "wideband SMA" (wSMA) will enable a substantial increase in the instantaneous bandwidth of the SMA to at least 16 GHz per sideband [1]. The upgrade will bring significant improvements in sensitivity and polarimetric performance, and will include provision for future upgrades.**

The new receiver system improves upon the previous SMA receiver optics, eliminating several room temperature receiver optics elements as well as the cooled receiver lenses, and introducing dual-polarized feedhorns which will greatly aid in polarimetric observations.

We present the methods used to design, optimize and tolerance the new receiver optics, starting from the imaging beam waveguide conditions and working through Gaussian beam analysis and simulation of the optical system with physical optics. We also describe the design of the new feeds for the wSMA system.

***Index Terms*—Antennas, feeds, quasioptics, radio astronomy, submillimeter astronomy.**

I. Introduction

THE Submillimeter Array (SMA) was conceived three decades ago as the world's first submillimeter interferometer capable of sub-arcsecond imaging in the frequency range from 200 to 700 GHz. Since it began full science operations in 2004 it has been continuously upgraded with new receiver cartridges, expanded intermediate frequency (IF) bandwidth, and augmented polarimetric and dual frequency observing modes. The next step in the development of the SMA is called the wideband Submillimeter Array (wSMA). The wSMA upgrade [1] will replace the original SMA cryostats, receivers, and receiver selection optics with all new systems, and incorporate a number of major upgrades to the backend IF signal transport and correlator systems. This will enable the wSMA to operate with 32 GHz instantaneous bandwidth per receiver polarization, and will form the basis of future development efforts.

The new wSMA receiver optics design is substantially different to that of the SMA receiver optics, with the replacement of room temperature and transmissive optics with cooled reflective optics, the elimination of quasioptical local oscillator (LO) injection, and major changes to the receiver selection mechanism. The new wSMA receiver optics are designed to couple to the SMA antennas' existing beam waveguide, with no changes to the antenna or beam waveguide.

In this paper, we present the design procedures and goals used in developing the wSMA receiver optics design, and the design procedures used to develop the new feeds for the receiver frontend modules. The initial design of the receiver optics was developed using the imaging beam waveguide conditions described by Chu [4], and then further refined using Gaussian beam analysis before carrying out physical optics simulations of the complete wSMA system from feed to sky using TICRA's GRASP software package, with the feeds being analyzed and optimized using both TICRA's CHAMP software and custom modal matching software. Custom software packages were developed in Python to analyze the GRASP and CHAMP outputs.

II. SMA Optical Design

The SMA antennas are of a bent Nasmyth design, fed by a beam waveguide that places the receiver feed horn aperture at an image of the antenna aperture (Fig. 1) [2]. This allows for frequency-independent illumination of the 6 m aperture from a

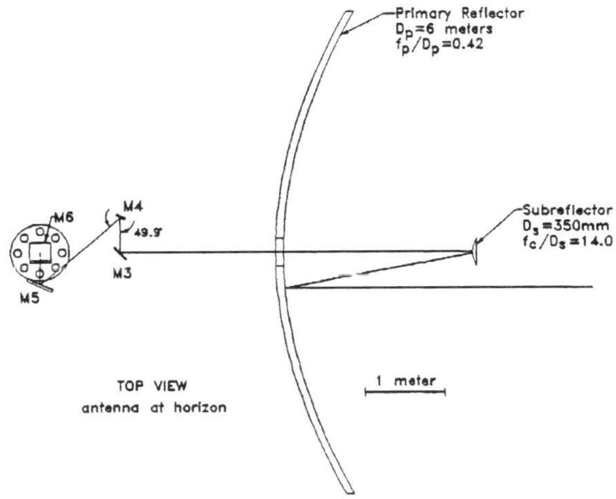

Fig. 1. Layout of the SMA antenna and beam waveguide, taken from [3]. M3 to M6 form the SMA beam waveguide, feeding the receiver system on the far left of the diagram.

The Submillimeter Array is a joint project between the Smithsonian Astrophysical Observatory and the Academia Sinica Institute of Astronomy and Astrophysics and is funded by the Smithsonian Institution and the Academia Sinica.

Paul K. Grimes, Scott N. Paine, Lingzhen Zeng, and Edward C.-Y. Tong are with the Center for Astrophysics | Harvard & Smithsonian, Cambridge, MA 02138, USA (e-mail: pgrimes@cfa.harvard.edu).

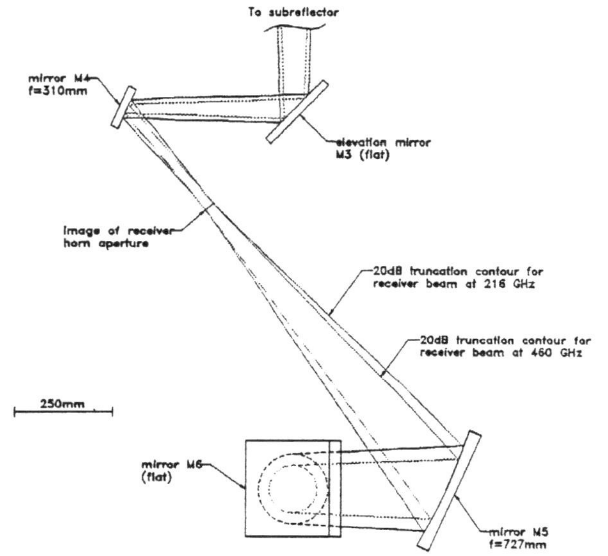

Fig. 2. Details of the SMA beam waveguide, taken from [3]. A movable calibration load and rotating quarter waveplates can be inserted at the image of the receiver horn aperture between M4 and M5.

feed with a frequency-independent aperture field distribution. In order to maximize the gain of the antennas, a 10 dB edge-taper is chosen.

The current SMA antennas use four common mirrors in the beam waveguide (M3/"tertiary" to M6, Fig. 2) to feed a moving wire grid and combiner mirror pair that split and direct orthogonal polarizations to two of four receiver optics inserts

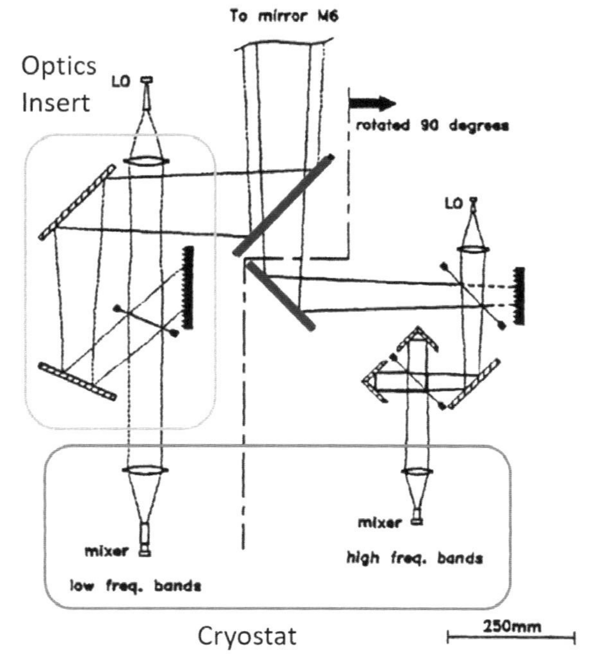

Fig. 3. The current SMA cryostat optical layout from [3], showing the moving grid and mirror receiver selectors (red), the room temperature LO injection optics in the Optics Insert (orange) and the cold receiver cartridge Teflon lens and feed horn. The right hand Optics Insert shows a Martin-Puplett diplexer LO injection scheme which was only used with the obsolete 690 GHz receivers.

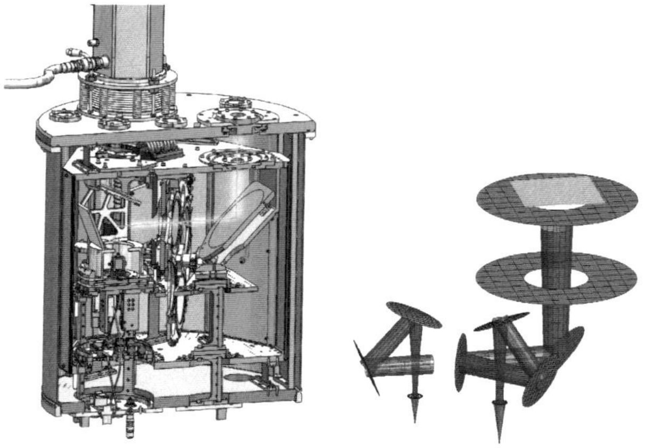

Fig. 4. (right) Cut-away CAD rendering of the wSMA cryostat, showing the cryostat structure and the optical layout (optical path highlighted in blue on the left image). (left) GRASP model of the wSMA cryostat optics, with the Gaussian beam from each feed shown in blue. Rendering courtesy of HPD Inc.

and receiver cartridges (Fig. 3). The optics inserts fold the beam before coupling LO power into the sky signal via a 99% reflective mesh. The beam then enters the receiver cartridge and is focused onto the feed horn by a Teflon lens cooled to ~80 K.

Careful alignment of the feed, receiver lens and optics inserts, and positioning of the combiner grid and mirrors is required in order to achieve coalignment of all receivers on the sky. This is of particular importance for polarimetric observations, as two orthogonally polarized signals travel through independent combiners, optical inserts, receiver lens and feeds, and misalignment can significantly affect the effective sensitivity of the antenna.

III. wSMA Optical Design

The new wSMA receivers will replace all of the optics from the wire grid to the receiver feeds. The beam from M6 enters the cryostat through a vacuum window and an IR filter on the 50 K radiation shield before being reflected through 90° by a plane mirror (Fig. 4). The beam then passes to a movable receiver "selector wheel", where it is either reflected to the low frequency receiver cartridge (210-270 GHz LO) or transmitted to the high frequency receiver cartridge (280-360 GHz LO). The selector wheel will also contain both a wire grid and a dichroic element to enable dual frequency observing in one of two possible polarization modes – either with one polarization per receiver cartridge or with both receiver cartridges dual polarized, with the latter mode processed at half the standard on-sky bandwidth. Both the fold mirror and selector wheel are mounted on the 50 K radiation shield top plate.

Each receiver cartridge will have a pair of "Receiver Optics" mirrors in a clam-shell arrangement that takes the incoming beam from the selector wheel and focuses it onto the dual-polarized receiver feed horn. The use of dual-polarized feeds ensures coalignment on the sky between the two orthogonally polarized beams.

The receiver optics and receiver front-end modules are mounted on a floating 4K cold-plate, which is aligned to the 50 K radiation shield top plate when the cartridge is inserted.

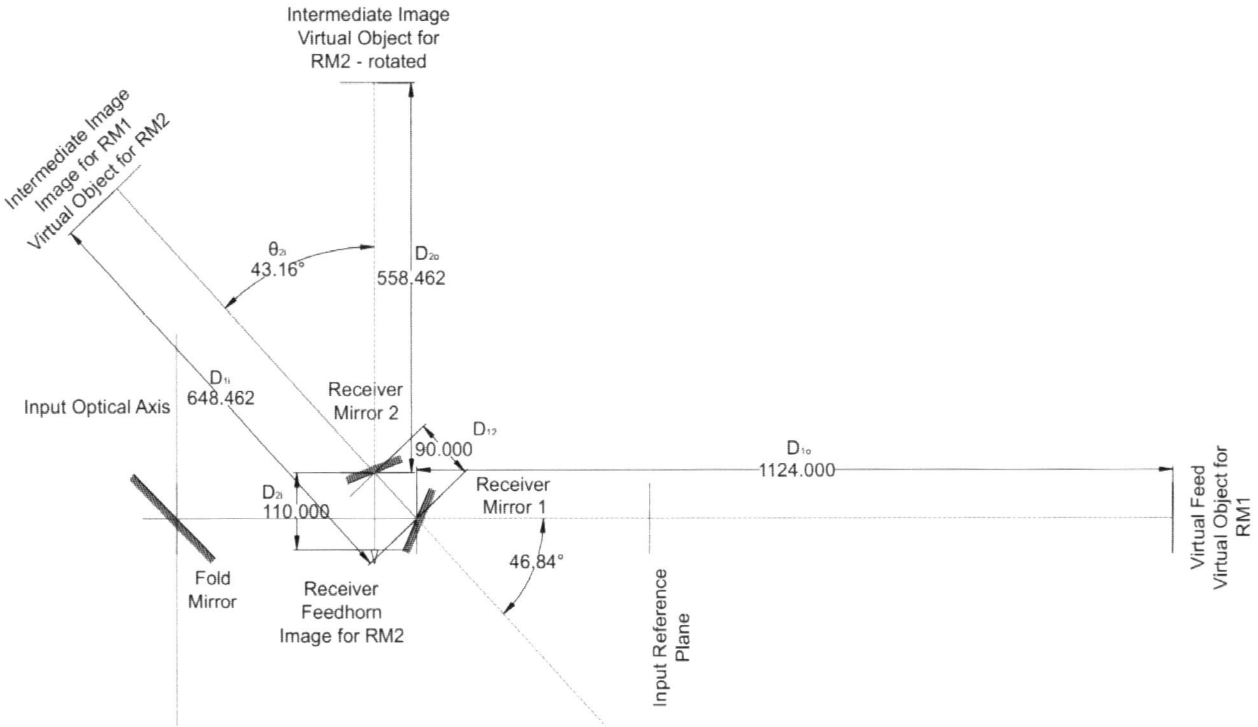

Fig. 5. The wSMA receiver optics layout, with dimensions used in the imaging calculations.

The 50 K top plate thus serves as a common mechanical alignment reference for all of the optical elements.

IV. Imaging Beam Waveguide Design

An imaging beam waveguide is a sequence of optical elements designed to provide frequency-independent matching between the input and output field distributions [4]. The frequency-independent Fresnel imaging equations provide a tool to design the optical system so that a frequency-independent match is obtained between input and output Gaussian beams (or any other field distribution).

The existing SMA beam waveguide was designed [3] to place a frequency independent virtual feed below each SMA receiver lens and feed, defined by its Gaussian beam radius and radius of phase curvature. The receiver lens reimages this virtual feed on the feedhorn aperture. Different receiver lens focal lengths and feedhorn apertures were chosen for each frequency band used in the SMA.

For the wSMA receivers, the fold mirror and selector wheel will direct the beam to each of the receiver cartridges, entering the receiver optics from the side and placing a virtual feed behind each of the receiver optics assemblies. Because we no longer need to fold the beam to inject the local-oscillator signal, the receiver optics intercept the beam further in front of the virtual feed. As the wSMA receivers cover a narrower total frequency range than those originally designed for the SMA, we adopt the same feed aperture and receiver optics design for both receiver bands, allowing receiver optics assemblies to be interchangeable between receiver bands.

Both mirrors in the clamshell pair are offset hyperboloidal mirrors. The first mirror reimages the virtual feed to an intermediate image in front of the first mirror and behind the second mirror. The second mirror then reimages the intermediate image to the feed aperture. The layout of the reimaging system is shown in Fig. 5.

As a convenience to allow existing SMA receiver feeds to be used with the wSMA receiver optics, we constrain the total magnification of the mirror pair (i.e. the scaling from the virtual feed beam radius to the feed beam radius) to be the same as that used for the SMA 300 GHz and 400 GHz receivers:

$$\text{Total Magnification: } m = m_1 \cdot m_2 = 0.11363636 \quad (1)$$

$$\text{Magnification of RM1: } m_1 = -\frac{d_{1i}}{d_{1o}} \quad (2)$$

$$\text{Magnification of RM2: } m_2 = -\frac{d_{2i}}{d_{2o}} \quad (3)$$

The position along the incoming beam at which the first mirror is placed is roughly determined by the overall physical layout of the optics, the need for compactness in order to fit within the cryostat, and the need to fit the cryostat within the receiver cabin.

The distances between the two mirrors and between the second mirror and the feed are chosen to avoid truncation of the beam by the edge of the second mirror or the feed, and to place the angles of reflection close to 45°. The beam widths are estimated from the single-mode Gaussian beam analysis of the beam waveguide described below, resulting in the following dimensions, as well as the relation between the image distance for RM1 and the object distance for RM2:

$$\text{Distance between RM2 and feed: } d_{2i} = 110 \text{ mm}$$
$$\text{Distance between mirrors: } d_{12} = 90 \text{ mm}$$

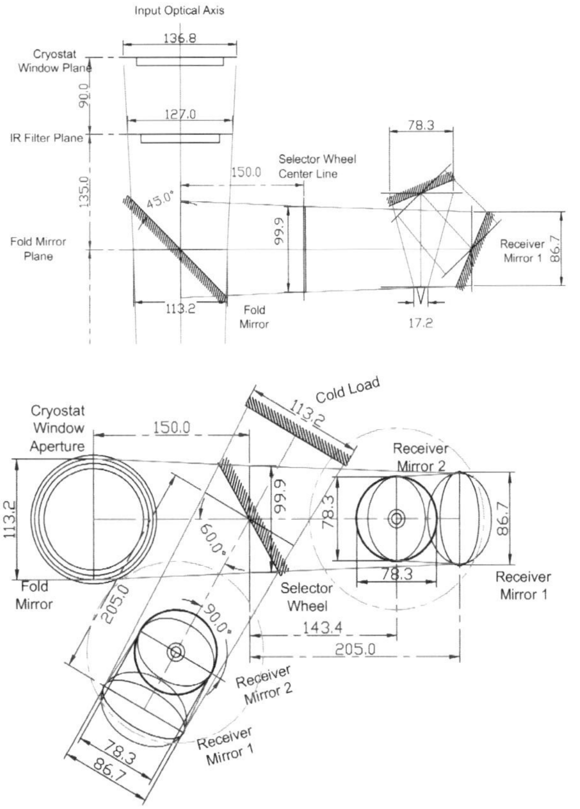

Fig. 6. The wSMA cryostat optical layout, from the side and above, showing the beam widths (blue dimensions) at each of the optical elements.

$$d_{2o} = d_{12} - d_{1i} \tag{4}$$

The position of the intermediate image is then determined by the above imaging equations subject to the constraint set by the chosen overall magnification, with the distance from the first mirror to the virtual feed

$$d_{1o} = 1124 \text{ mm}$$

as the driving parameter.

Solving the equations (1-4) for the intermediate image position, d_{1i}, we obtain:

$$d_{1i} = \frac{m \cdot d_{1o} \cdot d_{12}}{(d_{2i} + m \cdot d_{1o})} = 648.462 \text{ mm} \tag{5}$$

This in turn determines the input object for the second mirror and thus the individual magnifications of the two mirrors.

The precise angles of the mirrors are then set according to the Mizuguchi-Dragone condition:

$$\frac{(1 - m_1)}{m_1} \cdot \tan\theta_1 = (1 - m_2) \cdot \tan\theta_2 \tag{6}$$

which relates the individual reflection angles and magnifications of the two mirrors so that cross-polarization of the beam is minimized. By adding the constraint that the total reflection angle be 90°, the optical system is then fully determined.

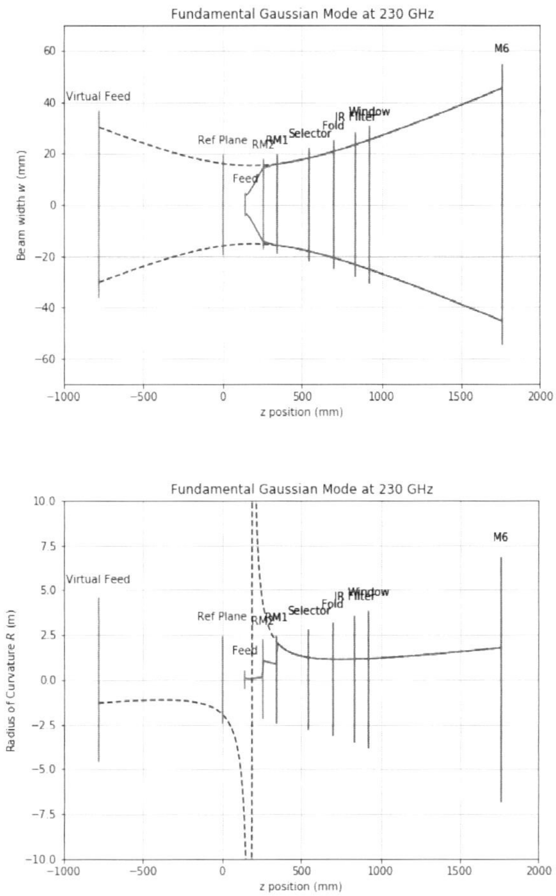

Fig. 7. Results of fundamental mode Gaussian beam analysis of the wSMA receiver optics. The beam from the SMA beam waveguide to the virtual feed is shown with dark blue dashes, while the beam as modified by the wSMA receiver optics is shown in red.

V. Gaussian Beam Analysis

Following the definition of the receiver mirrors using the imaging equations above, a single-moded Gaussian beam analysis of the optical system is used to determine the Gaussian beam radius and radius of phase curvature at the feed aperture, and to check that the mirrors in the system are at least $5w$ in diameter, in order to avoid excessive truncation [5].

The solution of the imaging equations and the Gaussian beam analysis were implemented in a SMath workbook and confirmed using the Gaussian beam propagation software originally developed for designing the SMA optics. The position of the first mirror d_{1o} and the overall optical layout had to be adjusted to give a positive radius of phase curvature at the feed aperture.

The evolution of the Gaussian beam parameters as they pass through the optical system are shown in Fig. 7.

VI. Feed Design

The feeds for the wSMA system are cylindrically symmetric corrugated horns feeding square waveguide inputs to the receiver orthomode transducers. The magnification of the

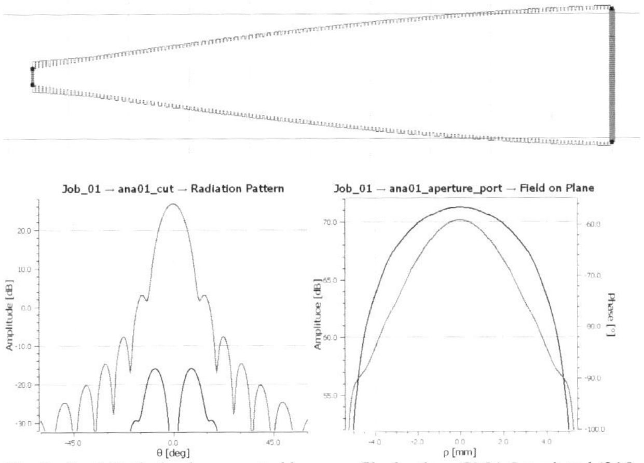

Fig. 8. (top) Optimized corrugated horn profile for the wSMA Low band (210-270 GHz) feed horn. (bottom left) Far-field E-plane radiation pattern at 230 GHz for the horn, showing co-polarized (red) and cross-polarized (black) fields. (bottom right) Aperture field amplitude (black) and phase (red) at 239 GHz for the co-polar E-plane field.

receiver optics was chosen to give the same Gaussian beam radius at the feed aperture as used by the SMA 300 GHz receiver's conical corrugated horn, and this horn serves as the starting point for the wSMA feed designs.

The Gaussian beam analysis above gives the Gaussian beam radius and radius of phase curvature at the horn aperture that must be matched by the wSMA feeds:

$$w = 3.4375 \text{ mm}$$
$$R = 159.64 \text{ mm}$$

Implementing such a horn using a traditional conical corrugated horn profile would require the horn to be prohibitively long, and so a curved horn profile is used.

Initial design of candidate horns were carried out in TICRA's CHAMP software, using modal-matching on CHAMP's built-in sinusoid and asymmetric sine-squared profiles, and variable-depth-slot mode converter designs, described in [6]. Output from the CHAMP software was analyzed in a Jupyter Python Notebook to obtain an estimate of the best fit fundamental Gaussian mode beam radius and radius of phase curvature for the aperture field.

The candidate horn designs were then further refined using modal-matching horn analysis software that has been designed to run massively parallel optimizations of horn designs, using a variant of Powell's minimization method [7] and a genetic algorithm. Additional goal functions that allow the complex aperture field mismatch to the fundamental Gaussian mode to be minimized were implemented for this project.

Finally, the optimized horn profiles were input back into TICRA CHAMP, and simulated as a complete feed horn consisting of the square-to-circular waveguide transition, mode-convertor, profiled corrugated horn and the cylindrically symmetric horn outside surface.

Final profiles, far-field beam patterns and aperture field distributions are shown for the Low band and High band feeds in Fig. 8 and Fig. 9 .

The output far-field beam patterns from CHAMP were then used in the GRASP physical optics simulations described in the next section. The GRASP calculated primary aperture illumination and efficiency of the complete wSMA optical system were then used to select the final horn designs.

VII. Physical Optics

Physical optics simulations of the complete SMA and wSMA optical system were implemented in TICRA's GRASP software. These physical optics simulations allowed a direct comparison to be made between the optical performance of the existing SMA receiver system and that of the proposed wSMA receiver system.

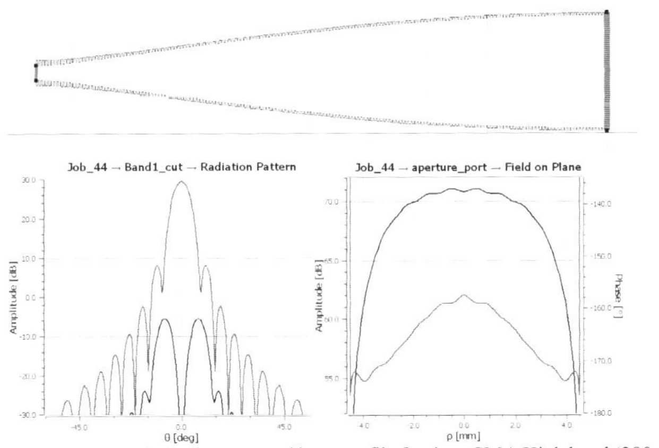

Fig. 9. (top) Optimized corrugated horn profile for the wSMA High band (280-360 GHz) feed horn. (bottom left) Far-field E-plane radiation pattern at 345 GHz for the horn, showing co-polarized (red) and cross-polarized (black) fields. (bottom right) Aperture field amplitude (black) and phase (red) at 345 GHz for the co-polar E-plane field.

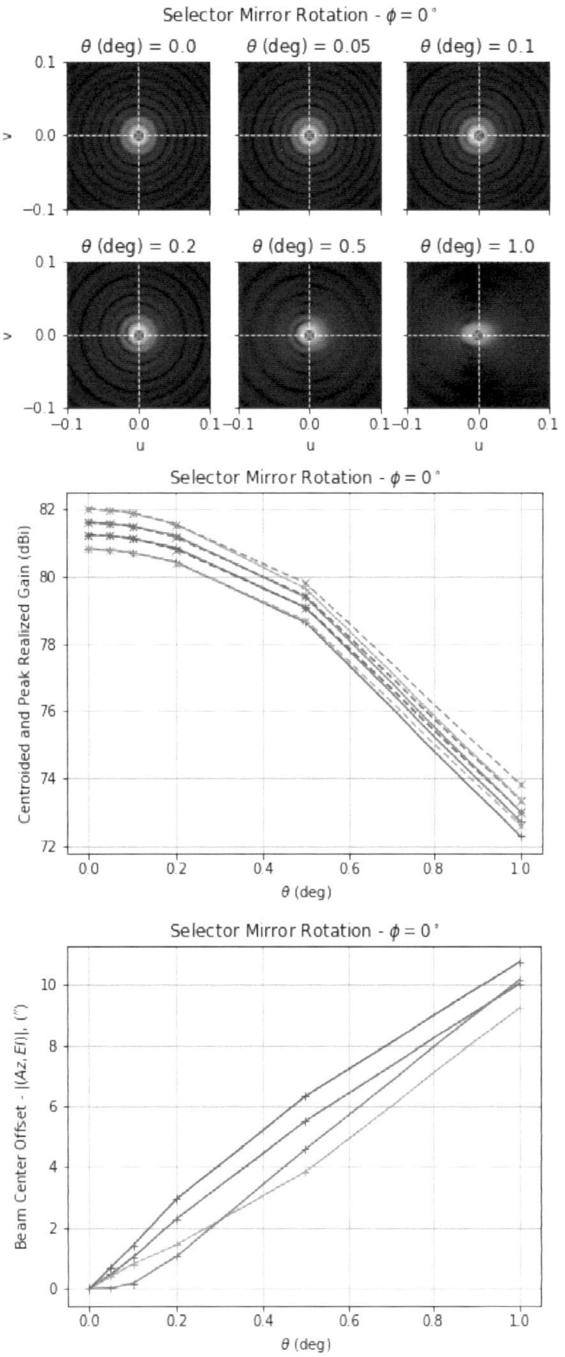

Fig. 11. Far field radiation pattern at 230 GHz (top), peak antenna gain (middle) and position of peak antenna gain (bottom) for 200, 210, 220 and 230 GHz farfield radiation patterns as the angle of the wSMA receiver selector wheel mirror is varied by up to 1°. These variations are used to set the tolerance on the angle of the selector wheel mirror in the wSMA cryostat design.

The GRASP model of the complete SMA optical system in shown in Fig. 10. The model implements physical optics methods for optical surfaces from feed to M3, and physical optics with correction from the physical theory of diffraction for the primary, secondary and key apertures in the optical system.

Key performance measures of the optics derived from the GRASP outputs include the peak gain of the antenna system,

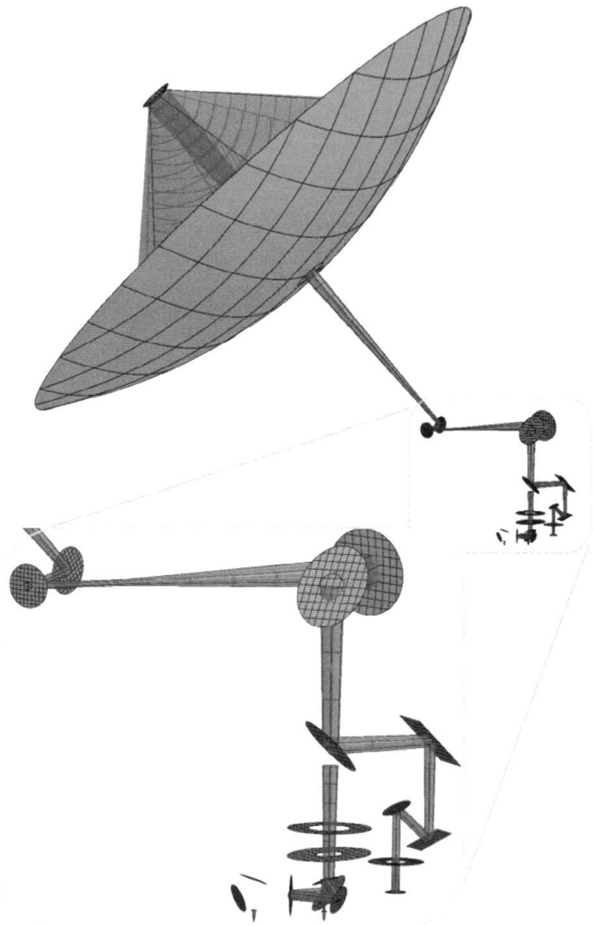

Fig. 10. GRASP model of the SMA antenna, beam waveguide and SMA and wSMA receiver optics, with the 230 GHz Gaussian beam radius shown in blue. An additional view of the wSMA receiver optics model is shown in Fig. 4.

the main beam efficiency (as measured by GRASP's spillover efficiency calculation), peak cross-polarization level, and coalignment between orthogonal polarizations.

In order to derive these measures from the GRASP output, a Python library was written to allow the output fields to be loaded as Numpy arrays [8]. Additional Python functions allowed the peak values and positions of the peaks in the fields to be determined.

The "python-graspfile" Python library for reading and processing GRASP output has been released under the Open Source MIT License, and can be obtained via and is available via the Python Package Index (PyPI) system or GitHub [9].

A. Tolerance Analysis

In order to determine appropriate tolerances for the placement of the receiver cryostat with respect to the telescope and for the positions and angles of the cooled cryostat and receiver optics, the position and angle of each element of the optics was varied individually.

The variation in peak gain and position of peak gain was determined for each variation. Tolerance thresholds were set at variations in position or angle that resulted in a reduction in peak gain of one percent.

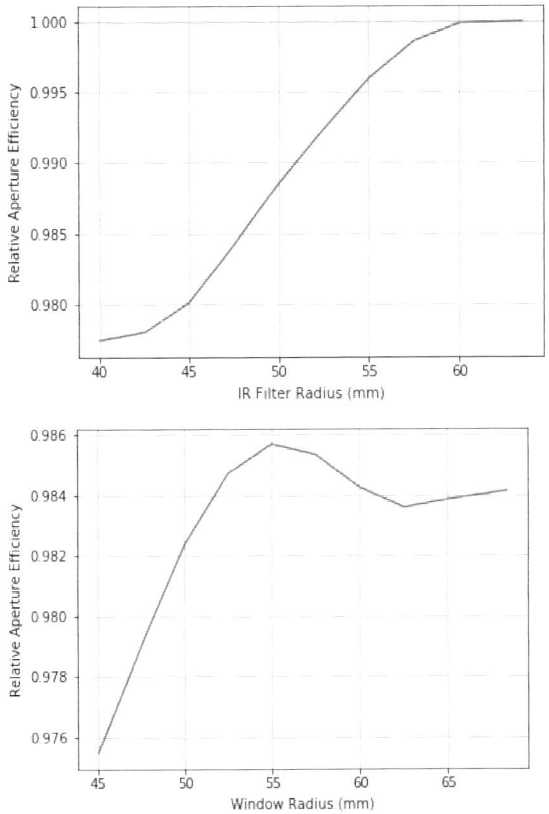

Fig. 12. Beam efficiency from the feed to M6 at 210 GHz as a function if IR Filter radius (top) and cryostat window radius (bottom) for the wSMA Low band receiver. The increase in efficiency at 55 mm window radius is not present at all frequencies within the band, and so the final clear aperture of 52.5 mm radius is a compromise that gives relatively good performance across the band of the Low Band receiver.

These simulations were carried out using the feed design from the SMA 300 GHz receiver. An example of this procedure for the angular tolerance on the wSMA receiver selector wheel mirror position is shown in Fig. 11.

B. *IR Aperture and Window Sizing*

The minimum size of the IR aperture, which acts as a cold aperture stop in the wSMA optical system, and cryostat window, was determined by running simulations with varying aperture radius, again using the SMA 300 GHz feed design. The minimum IR clear aperture was chosen so that the peak gain was reduced by somewhat less than one percent from the $r = 5w$ case at the lowest specified frequency for the wSMA Low Band receiver (210 GHz).

Once the IR aperture size was fixed, the minimum cryostat window clear aperture was determined by running simulations with fixed IR aperture and varying window size. The minimum window size was chosen to be the minimum size that did not appreciably truncate the beam with the IR aperture in place (total truncation are IR aperture and window < 1%). Beam efficiency results are shown in Fig. 12.

C. *Feed Evaluation*

Candidate feed designs (described above) were evaluated by comparing the GRASP output for simulations of wSMA receiver optics fed by each design (simulated in CHAMP) with the GRASP simulation output for the complete SMA 200 GHz or SMA 300 GHz receiver. Maximizing the peak gain and beam efficiency were the main criteria.

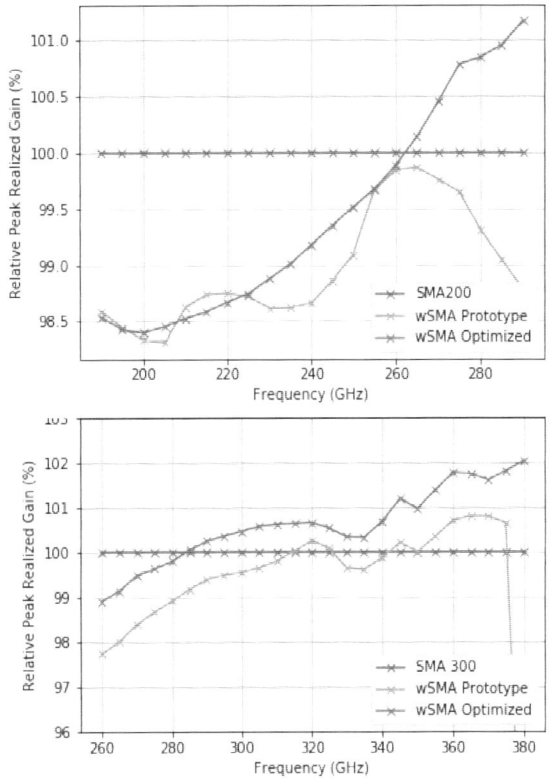

Fig. 13. Relative peak gain of an SMA antenna fed by the wSMA Low Band (top) and High Band (bottom) receivers using the initial prototype feed and the optimized feeds. The baselines for comparison are the existing SMA200 and SMA300 receivers.

The comparison of peak gain for the wSMA receiver fed by the prototype and optimized wSMA feeds with the SMA receivers is shown in Fig. 13. Although the aim was to achieve higher efficiency than the current SMA receivers, for the Low band receiver this would have required too large a window and IR aperture, and thus IR loading on the cryostat. Since the IR aperture forms a ~55 K cold stop in the optical system, the sensitivity of the Low band receiver should not be too badly affected. The High band receiver's narrower beam is not adversely affected by truncation at the IR aperture.

VIII. Conclusion

Two complete prototypes of the wSMA cryostat, receiver optics and receiver feeds are currently being built using the optical design presented in this paper. Following the integration of the wSMA cryostats and receivers, the performance of the optical system and the design will be measured in several ways.

Initial measurements will take place in the laboratory, using a beam scanner system similar to that described in [10]. This will verify that the optical system produces the correct beam, and that the beams of the two receivers are well aligned. Receiver performance measurements will also be carried out.

Following laboratory measurements, the prototype wSMA cryostat and receiver systems will be deployed in at least one SMA antenna at the SMA site on Maunakea. There, tests of the complete system will be made, including on-sky pointing and beam shape measurements, aperture efficiency, and holographic measurements of aperture field distribution of the primary mirror [11] .

References

[1] Paul K. Grimes, Raymond Blundell, Scott Paine, John Test, C-Y. Edward Tong, Robert W. Wilson, and Lingzhen Zeng, "The wSMA receivers – a new wideband receiver system for the Submillimeter Array", *28th Int. Symp. Space THz Tech.*, Cologne, Germany, 2017.

[2] Paul K. Grimes, Raymond Blundell, Patrick Leiker, Scott N. Paine, Edward C.-Y. Tong, Robert W. Wilson, and Lingzhen Zeng, "Receivers for the wideband Submillimeter Array", *31st Int. Symp. Space THz Tech.*, Tempe, Arizona, 2020.

[3] S. Paine, Cosmo Papa, R. Louie Leombruno, Xiaolei Zhang, and Raymond Blundell, "Beam Waveguide and Receiver Optics for the SMA," *5th Int. Symp. Space THz Tech.*, Ann Arbor, Michigan, 1994.

[4] Ta-Shing Chu, "An Imaging Beam Waveguide Feed", IEEE Trans. Antennas Propagat., AP-31 614, 1983.

[5] Paul Goldsmith, "Quasioptical Systems", IEEE Press. Piscataway, NJ, 1998.

[6] Christophe Granet and Graeme L. James, "Design of Corrugated Horns: A Primer", IEEE Ant. and Prop. Magazine, 47, 2, 2005.

[7] Lingzhen Zeng, C. Bennett, Chuss, and E. Wollack. "A Low Cross-Polarization Smooth-Walled Horn With Improved Bandwidth." IEEE Transactions on Antennas and Propagation, 58, 4, 2010.

[8] Stéfan van der Walt, S. Chris Colbert and Gaël Varoquaux. The NumPy Array: A Structure for Efficient Numerical Computation, Computing in Science & Engineering, 13, 22-30, 2011.

[9] Paul K. Grimes, "python-graspfile - Python module for reading output and configuration files", https://github.com/Smithsonian/python-graspfile, 2020.

[10] Christensen, Robert D, Ramprasad Rao, T. K Sridharan, and Edward Tong. "Vector Near-field Beam Scanner for the SMA." Proc. SPIE, 9153. 2014.

[11] Sridharan, Tirupati K, Masao Saito, Nimesh A Patel, and Robert D Christensen. "Holographic Surface Setting of the Sub-millimeter Array Antennas." Proc. SPIE, 5495, 2004.

Terahertz MgB_2 HEB mixers with a 13GHz gain bandwidth

Narendra Acharya, Evgenii Novoselov and Sergey Cherednichenko

Broad intermediate frequency (IF) bandwidth is a crucial feature for heterodyne receivers utilized in terahertz astronomy [1]. Whereas below 1.2 THz superconductor-insulator- superconductor (SIS) mixers have demonstrated an IF bandwidth in excess of 20GHz , at frequencies > 1.2THz superconducting hot-electron bolometer (HEB) mixers have not come to such level of performance yet.

We believe that utilizing MgB_2 ultra-thin [2] films the IF bandwidth for HEB mixers can be extended way above 20 GHz, hence satisfying the most demanding astronomical tasks. The key for a large IF bandwidth in HEB mixers is fast electron- phonon interaction and short phonon escape time from the superconducting film into the substrate. Recently we have shown that MgB_2 films as thin as 5nm can be made on SiC substrates, with a critical temperature of 30K. The minimum receiver noise temperature has been shown to be at 1000K (1.6THz) [3]. However, though the demonstrated IF bandwidth was 11GHz, we had some reasons to believe that this is far to be the limit.

By performing mixing experiments in a broadband cryogenic probe station we have been able to show that the gain bandwidth (3dB gain roll-off) in our MgB_2 HEB mixers is 13-14GHz, whereas the mixer noise temperature was nearly constant up to an intermediate frequency of 20GHz (Fig.1). These experiments were conducted at a rather low LO frequency of 100GHz, yet results have clearly shown that the full potential of MgB_2 HEB mixers is far not reached.

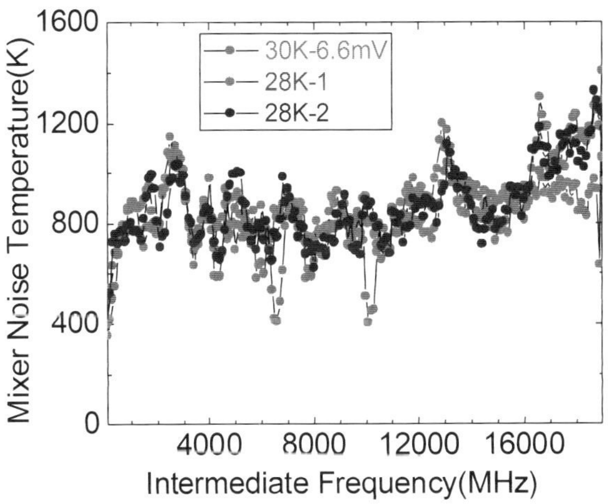

Fig. 1. The MgB_2 HEB mixer noise temperature spectrum.

In order to justify applicability of low LO frequency experiments for modeling the high LO frequency $\nu (h\nu > 2\Delta$, where Δ is the superconducting energy gap) we designed a set of experiments where MgB_2 HEB mixer performance (gain, noise, IF bandwidth) is studies vs LO frequency and operation temperature (which affects the superconducting energy gap Δ). Δ can be obtained from the measured kinetic inductance (from 5K and up to the critical temperature T_c).

References

[1] D. Farrah, K. E. Smith, D. Ardila, C. M. Bradford, M. Dipirro, C. Ferkinhoff, J. Glenn, P. Goldsmith, D. Leisawitz, T. Nikola, N. Rangwala, S. A. Rinehart, J. Staguhn, M. Zemcov, J. Zmuidzinas, and J. Bartlett, "Review: far-infrared instrumentation and technological development for the next decade," *J. Astron. Telesc. Instruments, Syst.*, vol. 5, no. 02, p. 1, 2019 , doi:10.1117/1.jatis.5.2.020901

[2] E. Novoselov and S. Cherednichenko, "Low noise terahertz MgB_2 hot-electron bolometer mixers with an 11 GHz bandwidth," *Appl. Phys. Lett.*, vol. 110, no. 3, 2017 , doi:10.1063/1.4974312.

[3] E. Novoselov and S. Cherednichenko, "Gain and Noise in THz MgB_2 Hot-Electron Bolometer Mixers with a 30-K Critical Temperature," *IEEE Trans. Terahertz Sci. Technol.*, vol. 7, no. 6, 2017 , doi:10.1109/TTHZ.2017.2760105.

[1] All authors work at the Department of Microtechnology and Nanoscience, Chalmers University of Technology, Fysikgrand 3, Gothenburg, Sweden, (e-mail: serguei@chalmers.se) (*Corresponding author Sergey Cherednichenko*).

Characterising the Effective Temperature of Hybrid Tunnel Junctions in THz SIS mixers

A. Traini, B-.K. Tan, J. Garrett, A. Khudchenko, R. Hesper, A. M. Baryshev, P. N. Dmitriev, V. P. Koshelets and G. Yassin

Abstract—**In this paper, we present a method to measure the effective physical temperature of the tunnel junction in a terahertz superconductor-insulator-superconductor (SIS) mixer when illuminated with a strong local oscillator (LO) power that results in local hot spot around the tunnel junction. We observed that the gap voltage of the pumped current-voltage (IV) curve is suppressed when the LO pumping level is increased, indicating that the junction physical temperature is increased beyond the mixer block temperature. We quantified this extra heating effect by recovering the effective junction temperature through comparing the gap voltage of the pumped IV curves measured at a fixed block temperature, with the unpumped IV curve measured at varying block temperatures. We found that the heat trapped in the tunnel junction can be as high as 1.7 K when the mixer stabilised at 3.3 K is pumped at only 21% of the gap current.**

Index Terms—**Terahertz SIS mixers, tunnel junctions, local oscillator heating effect.**

I. Introduction

SINCE the deployment of current Atacama Large Millimetre/sub-millimetre Array (ALMA), the Herschel Space Telescope, and the Stratospheric Observatory for Infrared Astronomy (SOFIA), the development of the next generation millimetre (mm) and sub-mm telescopes are now focusing on closing the 1–2 THz frequency gap. Observations at these frequency ranges are important for a wide range of scientific cases, including stellar formation at intermediate redshift universe and the Cosmic Infrared Background, where a host of spectral lines such as CO, NH^+ and other organic molecules can be detected [1]. Ground based observation in this frequency range is permitted through atmospheric transmission windows centred around 1.0, 1.3, and 1.5 THz [2]. To cover frequencies beyond these atmospheric windows, observation would need to be performed with high altitude or space observatories above the atmosphere.

This project is supported by the European Union's Horizon 2020 research and innovation programme under grant agreement No 730562 (RadioNet), the UK Science and Technology Facilities Council and in part by the Russian Science Foundation (Project No. 19-19-00618).

A. Traini, B-K. Tan and G. Yassin are with the Department of Physics (Astrophysics), University of Oxford, Denys Wilkinson Building, Keble Road, OX1 3RH, Oxford, UK. E-mail: alessandro.traini@physics.ox.ac.uk

J. Garrett is currently with the Harvard-Smithsonian Center for Astrophysics, 60 Garden St., Cambridge, MA 02138, USA., but conduct this work previously at the University of Oxford.

R. Hesper and A. M. Baryshev are with the Kapteyn Astronomical Institute, Groningen, 9747 AD The Netherlands.

A. Khudchenko, P. N. Dmitriev and V. P. Koshelets are with the Kotel'nikov Institute of Radio Engineering and Electronics, Russian Academy of Science, Moscow, 125009 Russia.

A. Khudchenco is also with Astro Space Center, Lebedev Physical Institute of Russian Academy of Science, 84/32 Profsoyuznaya st., Moscow, GSP-7, 117997, Russia.

Although it is possible to operate a heterodyne receiver using the hot electron bolometers (HEBs) in this frequency range, their deployment is unfavourable due to the relatively narrow intermediate frequency (IF) bandwidth. Superconductor-insulator-superconductor (SIS) mixers, on the other hands, can operate with much wider IF bandwidth, but have yet to demonstrate their feasibility above 1 THz. This is because of the limitation imposed by the commonly used niobium (Nb) technology. The performance of Nb tunnel junction tends to deteriorate when operating close to twice its gap frequency, around 1.3 THz. Furthermore, Nb based transmission line required to form the superconducting mixer circuit incur very high resistive losses above 0.7 THz.

In order to improve the sensitivity of SIS devices and extend the frequency coverage above 1.3 THz, Nb needs to be replaced with higher gap superconductors such as niobium nitride (NbN) or niobium titanium nitride (NbTiN). However, it has been shown that high quality full NbN or NbTiN tunnel junctions are difficult to fabricate. One solution is therefore to replace one of the Nb electrode with either NbN or NbTiN, forming a hybrid junction. However, these hybrid junctions working in conjunction with higher gap superconducting transmission lines could suffer from localised heating effect, due to quasiparticles trapping in the low gap superconductor sandwiched between high gap superconducting materials. This effect can be observed experimentally as shown in Fig. 1, where the superconducting gap voltage of the device is reduced notably when the mixer is injected with strong local oscillator (LO) power. This problem of heat trap in a tunnel junction has already been investigated theoretically [3], [4], [5], [6]. In this paper, we present an alternative method for measuring the effective temperature of the junction using only experimental data.

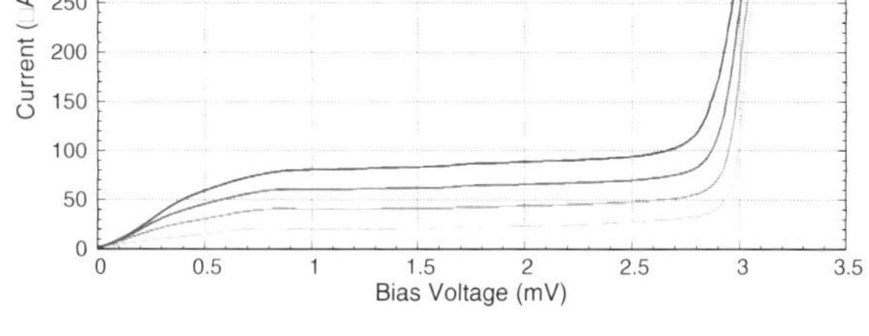

Fig. 1. Reduction of the gap voltage when the mixer, stabilised at a constant block temperature of 4 K, is injected with strong LO power.

II. Experimental Setup & Measurement Results

To investigate the amount of heat trapped in the tunnel junction when illuminated with strong LO power, we compare the gap voltage of a series of pumped current-voltage (IV) curves at varying pump levels and fixed mixer block temperature, with the unpumped IV curves at varying block temperature, since increasing the block temperature has the same effect of suppressing the gap voltage. The SIS mixer used in this experiment is a CHAMP+ (The Carbon Heterodyne Array of the MPIfR, Max Planck Institute for Radio Astronomy) mixer comprising niobium/aluminium-nitride/niobium-nitride (Nb/AlN/NbN) junctions deposited on NbTiN ground plane with aluminium (Al) as the top wiring layer [7].

The mixer was tested in the frequency range of 780–950 GHz, cooled with a pulsed-tube cooler cryostat, with all the optical components such as the beam splitter and the focusing mirrors mounted inside the cryostat to avoid the effect of water absorption [8]. The hot and cold loads required for Y-factor noise temperature measurements were also located inside the cryostat, with the respective physical temperature of 300 K and 4 K. The cryostat is also equipped with four temperature probes that were attached to the internal cold blackbody source, the mixer block, the cold plate and the second stage shield respectively. The cold plate and the second stage were further installed with resistors for raising the bath temperature above its minimum value. The current passing through the resistors is controlled by a proportional integral derivative (PID) controller loop made by Lake Shore®, to stabilise the mixer block temperature to a desired value, with accuracy of the order of mK.

Fig. 2 shows how we infer the effective temperature of the tunnel junctions. In this example, we first measured a pumped IV curve with the block temperature $T_{blk} = 4.0\,\mathrm{K}$, when illuminating with the LO set at at 831 GHz. We then measure a series of unpumped IV curves measured at varying T_{blk} =4.0–5.4 K, and look for the unpumped IV curve that have the same gap voltage as the pumped IV curve. As can be seen from Fig. 2, it is clear that the gap voltage of the measured pumped curve can only be matched with the unpumped curve where the block temperature is stabilised at 4.6 K. This indicates that

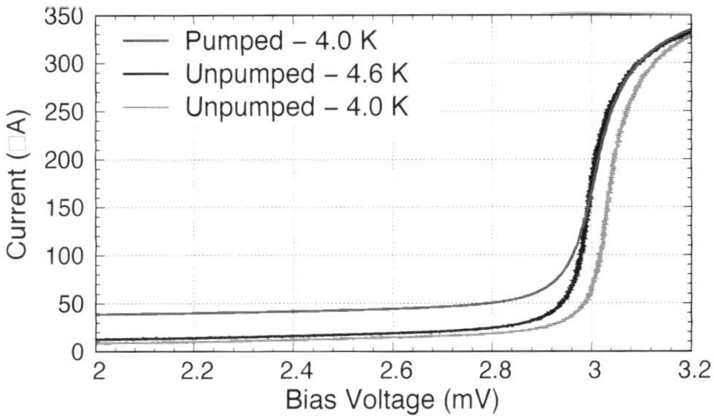

Fig. 2. Plot illustrating how we infer the effective temperature of the tunnel junction by matching the gap voltage of pumped and unpumped IV curves at different block temperature. The temperature shown in the legend indicates the mixer block temperature when the IV curve was taken.

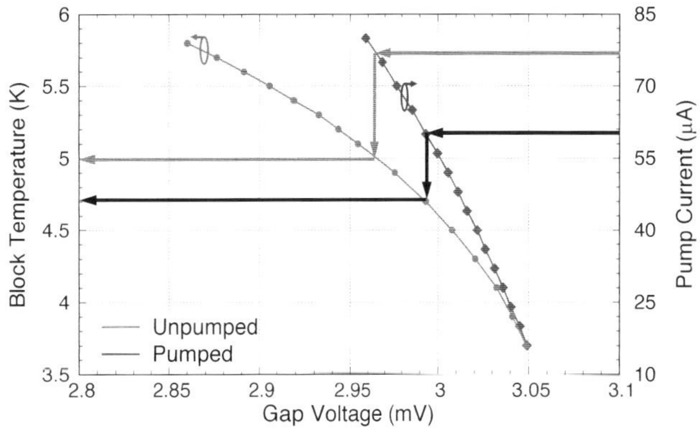

Fig. 3. Relation between the pump current and the effective tunnel junction temperature. The pumped IV curve was measured at a fixed temperature of 3.3 K.

there is an extra 0.6 K of heat trapped in the tunnel junction due to the LO.

We can also recover the effective junction temperature by correlating the mixer block temperature of the unpumped IV curves with the pump current, measured at the centre of the photon step of the pump curve, instead of the gap voltage. In this case, in addition to the series of unpumped IV curves measured at different mixer block temperature, we further measured a series of pumped IV curves with a fixed block temperature of 3.3 K but injected with LO at different power levels. Fig. 3 shows the relation between the two sets of data, and how we can easily estimate the effective tunnel junction temperature by reading the pump current. As shown in the black arrows, if the pump current is measured at say 57 μA, we can directly infer from the graph that the junction temperature has been heat up to 4.7 K. Similarly as shown in the grey arrows, if the pump current is measured at 72 μA, the effective temperature of the junction must have been increased to 5 K, closed to 1.7 K higher than the mixer block temperature.

III. Conclusion

We have characterised the effective temperature of the hybrid Nb/AlN/NbN tunnel junction fabricated between a NbTiN ground plane and an Al wiring layer, when pumped by a strong LO power operating near 1 THz regime. The increase of junction temperature due to the LO heating was estimated by matching the gap voltage of the pumped IV curve measured at a fixed block temperature, with the gap voltage of the unpumped curve measured at different block temperature. We have shown that the junction temperature can be increased by 1.7 K when the mixer is pumped at approximately 21% of the gap current, which will inevitably affect the performance of the mixer operating at such high frequency.

Acknowledgment

We thank Rik Elliott for his assistance in setting up the THz measurement system. The fabrication of the SIS mixer was carried out at the Kotel'nikov Institute of Radio Engineering and Electronics, Russian Academy of Science, within the framework of the state task (by USU 352529).

REFERENCES

[1] D. Rigopoulou, R. Laing, S. Withington, G. Magdis, S. Graves, J. Richer, and B. Ellison, "Science with ALMA Band 11 (1.0–1.6 THz)," *The Messenger*, vol. 153, pp. 35–37, 2013.

[2] H. Matsuo, S.-C. Shi, S. Paine, Q.-J. Yao, and Z.-H. Lin, "Terahertz atmospheric windows for high angular resolution terahertz astronomy from dome a," *arXiv preprint arXiv:1902.06398*, 2019.

[3] P. Dieleman, T. Klapwijk, S. Kovtonyuk, and H. van De Stadt, "Direct current heating in superconductor–insulator–superconductor tunnel devices for THz mixing applications," *Applied physics letters*, vol. 69, no. 3, pp. 418–420, 1996.

[4] B. Leone, B. Jackson, J. Gao, and T. Klapwijk, "Geometric heat trapping in niobium superconductor–insulator–superconductor mixers due to niobium titanium nitride leads," *Applied Physics Letters*, vol. 76, no. 6, pp. 780–782, 2000.

[5] B. Leone, J. Gao, T. Klapwijk, B. Jackson, W. Laauwen, and G. De Lange, "Hot electron effect in terahertz hybrid devices," *IEEE transactions on applied superconductivity*, vol. 11, no. 1, pp. 649–652, 2001.

[6] B. Tan, G. Yassin, P. Kittara, and J. Leech, "Measurement of electron-phonon interaction time of niobium using heating effect in SIS tunnel junction," in *Proceedings of 20th International Symposium Space THz Technology*, 2009.

[7] A. Khudchenko, A. M. Baryshev, K. I. Rudakov, P. M. Dmitriev, R. Hesper, L. de Jong, and V. P. Koshelets, "High-gap Nb-AlN-NbN SIS junctions for frequency band 790–950 GHz," *IEEE Transactions on Terahertz Science and Technology*, vol. 6, no. 1, pp. 127–132, 2016.

[8] B.-K. Tan, S. Mahashabde, A. Hector, G. Yassin, A. Khudchenko, R. Hesper, A. M. Baryshev, P. Dmitriev, K. Rudakov, and V. P. Koshelets, "Investigation of the performance of an SIS mixer with Nb-AlN-NbN tunnel junctions in the 780–950 GHz frequency band," in *Proceedings of 29th International Symposium Space THz Technology*, 2018.

A Novel Full Band Terahertz Frequency Quadrupler

F. Yang, *Member, IEEE*

***Abstract*—This paper presents designs of a novel full band quadrupler based on GaAs membrane technology. The measured quadrupler output power is more than -28 dBm at the full WR-1.0 band, and the measured peak power achieves above -13 dBm at 950 GHz with 7 dBm driven power.**

***Index Terms*—quadrupler, membrane substract, Schottky diode, submillimeter wave, terahertz.**

I. Introduction

Terahertz sources have attracted recent interest for terahertz measurement and radio astronomy applications [1], [2]. For such applications, it is important to have compact sources that produce enough power and own broadband performance.

Solid-state devices provide a compact and robust solution for the generation of terahertz signal. Transistor with smaller feature size have been demonstrated with maximum frequencies of oscillation (fmax) above 1 THz and practical circuit operation has been extended into the lower end of the terahertz (THz) frequency band. Although rapid advances have been made in amplifiers[3], especially high electron-mobility transistor (HEMT) [4] and heterojunction bipolar transistor (HBT) [5], Schottky diodes still provide high performance cost ratio and are widely used in terahertz frequency multiplier sources [6],[7]. For terahertz multiplier sources, cascading multipliers is a pragmatic approach. It typically consists of a chain of doublers and triplers, selected to yield the desired output frequency with a efficiency[8],[9]. The final efficiency of a multi-stage chain, as a result, is often on the order of a few percent or less [10]. Moreover, interstage mismatches in the chain can readily influence adjacent stages by pulling them from their optimum operating status and reducing efficiencies. Direct multiplication to a high-order harmonic greater than the third is desirable for higher frequency band, even with the challenges including proper termination of all intermediate harmonics (idlers). Also, a high-order harmonic multiplier brings the improvement on size, weight, power, and cost (SWaP-C).

This work presented here applies the architecture to implement an integrated frequency quadrupler fabricated on GaAs membrane substrate. This Quadrupler features 2 anodes in series configuration monolithically is integrated on a GaAs membrane substrate and connected to a split waveguide-block by metallic beam-leads. As we know. It is the first quadrupler working at the full WR-1.0 Band.

II. Quadrupler Development

A. Circuit topology

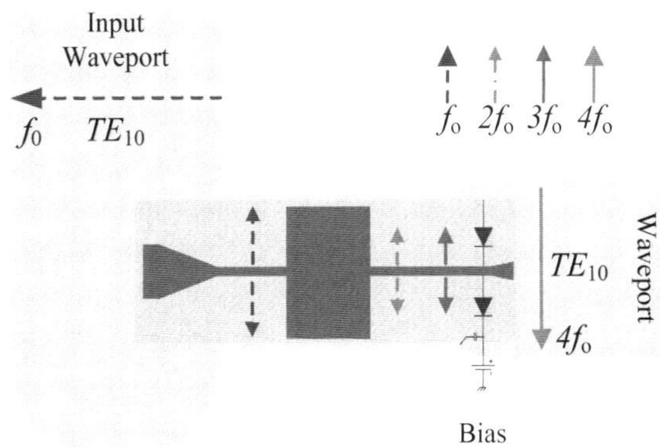

Fig.1. Block diagram of the balanced quadrupler.
f_o: the input fundamental frequency,
$2f_o$: the second idle harmonic frequency,
$3f_o$: the third idle harmonic frequency,
$4f_o$: the output fourth harmonic frequency.

Fig. 1 illustrates the basic architecture of the quadrupler circuit developed in this work. The diodes pair is placed in the output waveguide, and the two diode branches are antiparallel connected to the input fundamental signal on a TEM mode and series connected to the output 4th harmonic output signal on a TE mode. Usually, only even order harmonics are generated due to the balanced configuration. Considering small asymmetries are introduced by the physical structure of the Schottky diodes, there is still low level 3rd harmonic in the circuit. The low-high impedance filter are placed within the channel waveguide between the input and output waveguides which is used to match the diodes and prevent the 2nd and 3rd harmonics (idler frequencies) from leaking into the input waveguide. 2nd and 3rd harmonics are cut off by the output waveguide and trapped in the diodes circuit. Also, the dimension of the chip channel and the suspended microstrip line are optimized to trap the idler frequencies in the circuit.

B. Design Methodology

The methodology used for the design uses a combination of

This work was supported in part by the National Key R&D Program of China (Grant No.2018YFF0109302), the National Natural Science Foundation of China (Grant No.11573007).

F. Yang is with the State Key Lab of Millimeter Waves, Southeast University, Nanjing 210096, China (e-mail: yangfei@seu.edu.cn).

non-linear circuit simulations (Agilent ADS) and 3D electromagnetic simulations (Ansoft HFSS) that is based on the methodology presented in [8],[9]. To start with, a standard diode model customized with in-house parameters was implemented together with its close 3D passive environment (diode cell), plus ideal input and output matching networks. During this simulation, the anode zero junction capacitance and the diode cell geometry, including anodes positions, input channel width and input stub length, were optimized using the harmonic balance and optimization routines of ADS in order to reach optimum power efficiency. Then, each transition in the circuit including step impedances, coupling probes and waveguide transitions were simulated within the 3-D electromagnetic environment with appropriate boundaries, waveports assignment and de-embedding planes. The simulation outcomes (2-D S-parameter matrices and their attenuations, impedances and permittivity values at central uided frequencies) were used in a global non-linear optimization. During this second optimization step, ideal $\lambda/4$ initial lengths were defined in the step-impedance filter and matching network in order to converge to optimum lengths of the circuit transmission lines. Several iterations were necessary to define the appropriate dimensions of the position of the input back-short, the diodes geometry, the input probe and the output probe near the diodes in order to reach a wide impedance matching over the input frequency band. Finally, the values found during the previous steps were feed back in HFSS to build the full circuit structure as shown in Fig. 1, and the circuit was finally simulated with the harmonic balance routine to check the coupling efficiency, global efficiency and output power. The electrical parameters of the Schottky diode model considered in the simulations are a series resistance Rs =30 Ω, an intrinsic zero voltage junction capacitance of Cjo = 1.6 fF, an ideality factor =1.3 and a built-in potential Vbi = 0.8 V. This design was optimized for 5.5 mW of input power and -1 V bias included.

III. Fabrication and Assembling

A. Circuit fabrication

As described in the previous section the fabricated quadrupler circuit features two diodes integrated on a 5 um thick GaAs membrane. The membrane epitaxial layers were grown by MBE technique, fine structures, like the anodes and the air-bridges were using the electron beam lithography for its highest precision. The anodes are about 0.6 um × 0.6 um, the mesas are about 7.6 um, and the doping of the epilayer is 5×10^{17}. The gap between the edges of the anode and the ohmic contact was set to about 0.7 um. The chip is inserted between the input and the output waveguides in a channel of 93 um × 52 um cross section and approximately 232 um long.

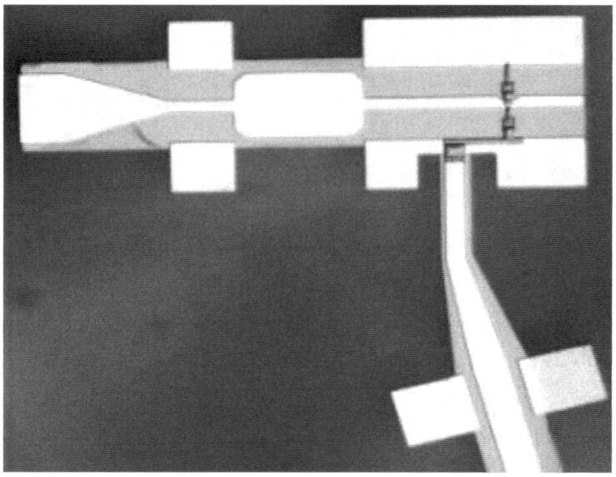

Fig.2. 750–1100 GHz circuit with beam-lead

B. Circuit assembling

The beam leads are thin gold layers that allow suspending the circuit in the channel without using conductive glue or golden bonding. The membrane circuit has been assembled into one of two halves and the beam leads are thermo-compressed on the block. The introduction of beam provides more precise RF and DC grounding and takes the significant improvement during the chip handling and assembling into the blocks. Protruding part of the chip is the DC line for biasing. It can also be used to grab the chip during the chip assembling into the blocks.

IV. Measurement

A. DC characteristic

Before electrical assembly, I-V on-wafer measurement have been performed, giving excellent diode I-V curve and ΔV, with the ΔV defined as ΔV = V (I=100 μA) - V (I=10 μA) = 72 mV. The extracted electrical parameters of the Schottky diode from the curve fitting are: series resistance R_s = 18 ohms, an ideality factor η = 1.27, and a built-in potential of V_{bi} = 0.78 V. The series resistance here is larger than the value mentioned in section II, which is partly due to the anode diameter after the fabrication is 0.8 um × 0.8 um rather than the value of 0.6 um × 0.6 um in simulation procedure.

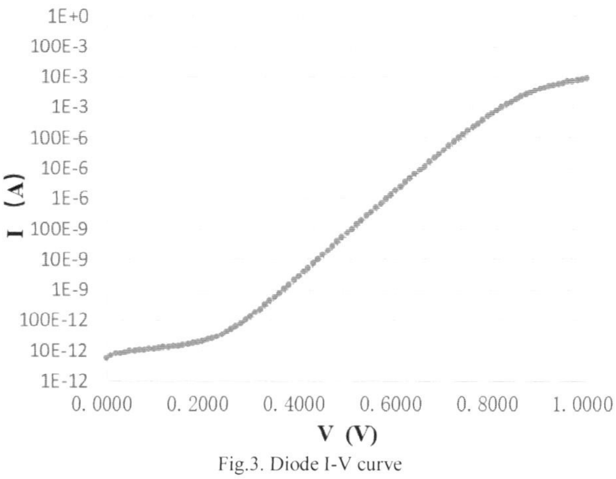

Fig.3. Diode I-V curve

B. *RF characteristic*

Another test measures the output power performance of the full band quadrupler. As illustrated in Fig.3, the power measurement gives -28 dBm minimum power across the 750-1100 GHz band, with -13dBm peak output power at 950 GHz. The measurement has very similar trend and a good fit to the design simulation results[11].

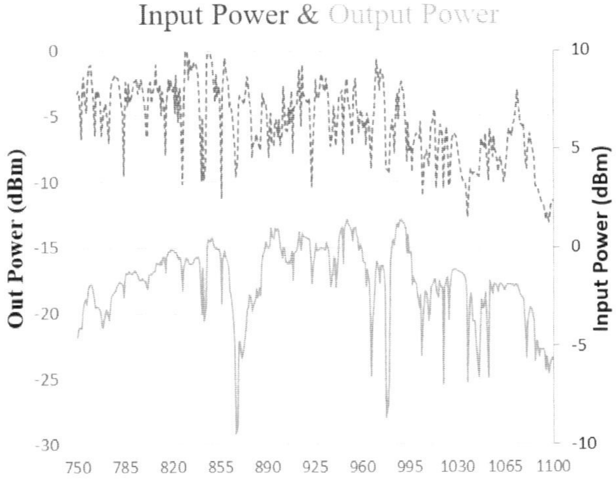

Fig.4. 750–1100 GHz circuit performance

C. *Comparison to Previously Published Results*

Table I provides a comparison of the current results against published quadruplers.

TABLE I
COMPARE WITH SIMILAR WORKS

Reference /Year	Technology	Bandwidth	Output Power
[12]/2014	0.13 μm SiGe HBT	50 - 75 GHz	-1 dBm
[13]/2014	0.12 μm SiGe BiCMOS	70 - 110 GHz	-1.5 to 2 dBm
[14]/2020	0.25 InP DHBT	110 - 130 GHz	5 to 7 dBm
[15]/2014	quasi-vertical diode	140 - 170 GHz	12 to 18 dBm
[16]/2013	45 nm SOI CMOS	275 - 285 GHz	-7 dBm
[17]/2015	45 nm SOI CMOS	390 - 440 GHz	-10 dBm
[18]/2012	35 nm mHEMT	435 - 480 GHz	-14.3 dBm
[19]/2017	Schottky barrier diode	322 - 345 GHz	0 dBm
[20]/1984	Schottky barrier diode	310 - 345 GHz	-3 dBm
[21]/1979	Schottky barrier diode	589 GHz	-28 dBm
This work	Schottky barrier diode	750 - 1100 GHz	-28 to -13 dBm

V. Conclusion

The Schottky diodes on GaAs membrane were fabricated with a process entirely based on electron-beam lithography. The performance shows its potential to be used as the extended parts for terahertz vector network analyzer.

Acknowledgment

The development of Schottky devices for terahertz measurement instrument is designed and fabricated at Nanjing. The authors would like to thank Dr. Niu B. for giving his kind suggestion for the fabrication process. Also, the author would like to acknowledge Dr. Deng J.Q. and Dr. Jia D.H. for their admirable patient during the THz-MIC mounting and test.

References

[1] J. J. Lynch, et al. "Coded aperture subreflector array for high resolution radar imaging." *SPIE Defense + Security International Society for Optics and Photonics*, 2017.

[2] I. Mehdi and P. Siegel, "THz semiconductor-based front-end receiver technology for space applications," *in IEEE Radio Wireless Conf.*, Atlanta, GA, Sep. 2004, pp. 127–130.

[3] L. A. Samoska, "An Overview of Solid-State Integrated Circuit Amplifiers in the Submillimeter-Wave and THz Regime," *in IEEE Transactions on Terahertz Science and Technology*, vol. 1, no. 1, pp. 9-24, Sept. 2011, doi: 10.1109/TTHZ.2011.2159558.

[4] X. Mei et al., "First demonstration of amplification at 1THz using a 25-nm InP high electron mobility transistor process," *in IEEE Electron Device Lett.*, vol. 36, no. 4, pp. 229–327, Apr. 2015.

[5] B. Heinemann *et al.*, "SiGe HBT with fx/fmax of 505 GHz/720 GHz," *2016 IEEE International Electron Devices Meeting (IEDM)*, San Francisco, CA, 2016, pp. 3.1.1-3.1.4, doi: 10.1109/IEDM.2016.7838335.

[6] I. Mehdi, E. Schlecht, G. Chattopadhyay, and P. H. Siegel, "THz local oscillator sources: Performance and capabilities," *Proc. SPIE*, vol.4855, pp. 435–446, 2002.

[7] A. Maestrini et al., "A Frequency-Multiplied Source With More Than 1 mW of Power Across the 840–900-GHz Band," *in IEEE Transactions on Microwave Theory and Techniques*, vol. 58, no. 7, pp. 1925-1932, July 2010, doi: 10.1109/TMTT.2010.2050171.

[8] A. Maestrini et al., "1200GHz and 600GHz Schottky receivers for JUICE-SWI". *27th International Symposium on Space Terahertz Technology.*

[9] A. Maestrini, B. Thomas, H. Wang, et al. "Schottky diode-based terahertz frequency multipliers and mixers". *in Comptes rendus - Physique*, 2010, vol. 11, no. 7, pp.480-495.

[10] I. Mehdi, J. V. Siles, C. Lee and E. Schlecht, "THz Diode Technology: Status, Prospects, and Applications," *in Proceedings of the IEEE*, vol. 105, no. 6, pp. 990-1007, June 2017, doi: 10.1109/JPROC.2017.2650235.

[11] F. Yang, "A 900GHz Broadband Balanced Frequency Quadrupler," *International Symposium on Space Terahertz Technology*, 2019.

[12] S. Yuan and H. Schumacher, "Novel frequency quadrupler design covering the entire V-Band in 0.13-μm SiGe process," 2014 IEEE 14th Topical Meeting on Silicon Monolithic Integrated Circuits in Rf Systems, Newport Beach, CA, 2014, pp. 101-103, doi: 10.1109/SiRF.2014.6828509

[13] B. Ku, H. Chung and G. M. Rebeiz, "A Milliwatt-Level 70–110 GHz Frequency Quadrupler With >30 dBc Harmonic Rejection," in IEEE Transactions on Microwave Theory and Techniques, vol. 68, no. 5, pp. 1697-1705, May 2020, doi: 10.1109/TMTT.2020.2967390

[14]M. Bao, R. Kozhuharov, J. Chen and H. Zirath, "A D-Band Keyable High Efficiency Frequency Quadrupler," in IEEE Microwave and Wireless Components Letters, vol. 24, no. 11, pp. 793-795, Nov. 2014, doi: 10.1109/LMWC.2014.2350694

[15] N. Alijabbari, M. F. Bauwens and R. M. Weikle, "160 GHz Balanced Frequency Quadruplers Based on Quasi-Vertical Schottky Varactors Integrated on Micromachined Silicon," in IEEE Transactions on Terahertz Science and Technology, vol. 4, no. 6, pp. 678-685, Nov. 2014, doi: 10.1109/TTHZ.2014.2360983

[16] M. Abbasi and D. S. Ricketts, "275–285 GHz balanced frequency quadrupler chain in 45 nm SOI CMOS," in Electronics Letters, vol. 51, no. 18, pp. 1424-1426, 3 9 2015, doi: 10.1049/el.2015.2100.

[17] F. Golcuk, O. D. Gurbuz and G. M. Rebeiz, "A 0.39–0.44 THz 2x4 Amplifier-Quadrupler Array With Peak EIRP of 3–4 dBm," in IEEE Transactions on Microwave Theory and Techniques, vol. 61, no. 12, pp. 4483-4491, Dec. 2013, doi: 10.1109/TMTT.2013.2287493.

[18] I. Kallfass, A. Tessmann, H. Massler, S. Wagner and A. Leuther, "A 480 GHz active frequency multiplier-by-four SMMIC," 2012 IEEE/MTT-S International Microwave Symposium Digest, Montreal, QC, 2012, pp. 1-3, doi: 10.1109/MWSYM.2012.6258409.

[19] J. Jun et al., "Single-Stage Frequency Quadrupler With SBD and Four-Octave LPF at 335 GHz," in IEEE Transactions on Terahertz Science and Technology, vol. 7, no. 4, pp. 446-454, July 2017, doi: 10.1109/TTHZ.2017.2709720.

[20]J. W. Archer, "A Novel Quasi-Optical Frequency Multiplier Design for Millimeter and Submillimeter Wavelengths," in IEEE Transactions on Microwave Theory and Techniques, vol. 32, no. 4, pp. 421-427, Apr. 1984, doi: 10.1109/TMTT.1984.1132693.

[21] T. Takada and M. Ohmori, "Frequency Triplers and Quadruplers with GaAs Schottky-Barrier Diodes at 450 and 600 GHz," in IEEE Transactions on Microwave Theory and Techniques, vol. 27, no. 5, pp. 519-523, May 1979, doi:

10.1109/TMTT.1979.1129660.

F. Yang (M'17) was born in Jiangsu, China, in 1978. He received the M.E. and Ph.D. degrees from Southeast University, Nanjing, China, in 2004 and 2007, respectively.

From 2008 to now, He works in the School of Information Science and Engineering, Southeast University, Nanjing, China.

His research interests include millimeter-wave low noise devices and submillimeter devices.

Thermal Transport in Graphene-based Hot Electron Bolometers with Different Electrode Contacts

W. Miao, F.M. Li, H. Gao, Z. Wang, W. Zhang, Y. Ren, K.M. Zhou, S.C. Shi, C. Yu, Z.Z. He, Q.B. Liu, and Z.H. Feng

We report on the experimental study of thermal transport of hot electrons in graphene-based hot electron bolometers (i.e., graphene microbridges) with different contacts. In the experiment, the temperature of hot electrons in a graphene microbridge is probed based on Johnson noise thermometry. We measured the thermal conductance of two samples at different bath temperatures, one with its electrodes of normal metal Au and the other with superconducting Nb electrodes. We found that thermal coupling through electron diffusion is dominant in the graphene microbridge with Au electrodes at temperatures below 9 K, and the measured thermal conductance follows the Wiedemann-Franz law, i.e., with linear temperature dependence. The unexpected thermal coupling through electron diffusion limits the sensitivity of graphene-based hot electron bolometers [1]. Hence we adopted superconducting Nb electrodes to prevent out-diffusion of hot electrons. It has been found that with the aid of superconducting electrodes, hot electron energy can be well confined within the microbridge, and the thermal conductance is reduced by nearly two orders at 0.3 K.

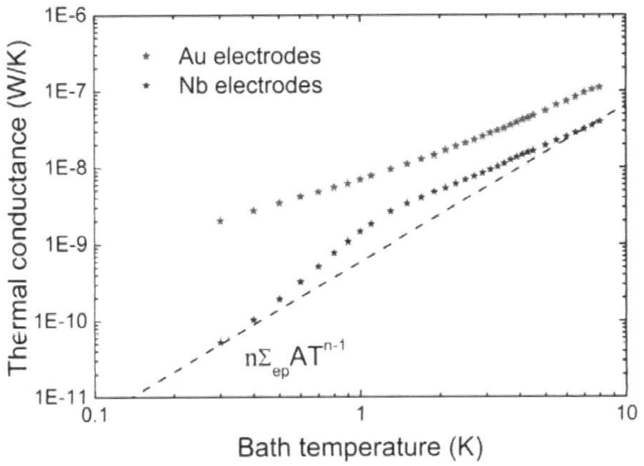

Fig. 1. Measured thermal conductance of the graphene-based hot electron bolometers with Au contacts and with Nb contacts at different bath temperatures.

Purple Mountain Observatory, CAS, Nanjing 210033, China.

REFERENCES

[1] W. Miao, H. Gao, Z. Wang, W. Zhang, Y. Ren, K.M. Zhou, S.C. Shi, C. Yu, Z.Z. He, Q.B. Liu, and Z.H. Feng, "A Graphene-Based Terahertz Hot Electron Bolometer with Johnson Noise Readout," *J. Low Temp. Phys.*, vol. 193, no. 3-4, pp.387 –392, Nov. 2018.

NOTES:

Development of a 350-GHz Dual-Polarization On-Chip Spectrometer

Jing Li[1,2], B.W. Fan[1,3], Z.H. Peng[1,3], J.Q. Ding[1], and S.C. Shi[1,2]

Spectral imaging observations provide information on the composition and physical conditions of observed astronomical sources as well as their redshifts. Broadband imaging spectrometers are therefore playing an increasingly important role in terahertz astronomy. While different spectrometer techniques (e.g., FTS and grating) have been explored, on-chip integrated superconducting spectrometers such as SuperSpec [1] and DESHIMA [2], which are much more compact, are emerging as powerful imaging spectrometers in the THz regime (loosely defined as 0.1-10 THz). Such an imaging spectrometer is based on a superconducting filter bank of low loss and high selectivity that splits a broadband signal into different frequency channels with a typical frequency resolution of 100-1000. Note that the broadband signal is firstly received by a broadband planar antenna and then transmitted to the superconducting filter bank. Each filter in the filter bank is followed by a microwave kinetic inductance detector for power detection. Obviously the detection sensitivity is defined by both the MKIDs and the filters. Furthermore, imaging is achieved by simply adding more pixels in a single chip, with each pixel providing an independent spectrum.

Imaging spectrometers of this kind are still in their infancy, but they have received widespread attention. In this paper, we introduce a 350-GHz on-chip integrated superconducting spectrometer, which is developed for technical demonstration on the POST submillimeter telescope [3]. Based on aluminum (Al) MKIDs incorporating a Nb/SiO_2/Nb microstrip-line filter bank, we are developing a 350-GHz on-chip integrated superconducting spectrometer with a frequency resolution of over 200. Detailed filter simulation and dark noise measurement results are presented.

[1] Purple Mountain Observatory, Chinese Academy of Sciences, Nanjing, China

[2] Key Laboratory of Radio Astronomy, Chinese Academy of Sciences, Nanjing, China

[3] University of Science and Technology of China, Hefei 230026, China

REFERENCES

[1] P. S. Barry et al., *Proc. SPIE* 8452, 84522F-1 (2012), DOI: 10.1117/12.927089.

[2] A. Endo et al., *Proc. SPIE* 8452, 84520X (2012), DOI: 10.1117/12.925637.

[3] S. P. Huang et al., *Proc. AP-RASC*, 412 (2004).

NOTES:

Follow-up experiments of the gain and noise IF bandwidth for a Ni-NbN HEBM

Yoshihisa Irimajiri, Akira Kawakami

We are developing an HEBM at 2 THz for atmospheric research to measure emission line spectrum of OH at 1.83 THz and O-atom at 2.06 THz from space. We have reported widening of the gain IF bandwidth of the HEBM using a newly proposed HEBM device which uses a magnetic thin film to suppress the superconductivity under the electrode which is called a Ni-NbN HEBM [1, 2]. The gain IF bandwidth was measured to be ~6.9 GHz. The gain IF bandwidth was measured by measuring a signal to noise ratio (SNR) of the beat note of the THz signal generated by an amplifier/multiplier chain (AMC) source with a fixed frequency and an output power. To measure the SNR of the beat note, the LO (AMC source) frequency was changed to change the beat note frequency. When the LO frequency is changed, the LO power is also changed which affects the operating conditions of the mixer, therefore, the LO power was adjusted to keep to the bias current of the mixer same at each LO frequency.

In this experiments, however, we did not much care about the saturation problem of the received signal. That is, if the input THz power is too much, the IF output of the HEBM will saturate and lead to incorrect results. The IF bandwidth was measured for the input THz power without attenuation of the input power (i.e., full power of ~9 μW) and for the 15 % of the power. The result shows that for the full power input, the IF response was almost flat with the SNR of ~53 dB until around 8 GHz. For the 15 % of the power input, the IF level gradually decreased and result shows a reliable bandwidth. From this experiment, we can conclude that our previously measured gain IF bandwidth was not affected by such a saturation effect.

To confirm the measured result of the IF bandwidth, we have also measured the noise IF bandwidth for the normal NbN and the Ni-NbN HEBM. The receiver noise temperature was measured by the conventional Y-factor method using a LNA and a room temperature amplifier with a frequency range of 1-12 GHz. A cryogenic amplifier/multiplier chain (AMC) source at 1.94 THz was used as a local oscillator (LO). The two triplers of the AMC source were cooled to 50 K using a mechanical cooler [3]. Although we need more confirmation experiments, the result (Fig. 1) shows the wider noise IF bandwidth for the Ni-NbN HEBM of ~7 GHz than that for the NbN HEBM of ~4.7 GHz. The measured noise IF bandwidth is almost the same as the gain IF bandwidth.

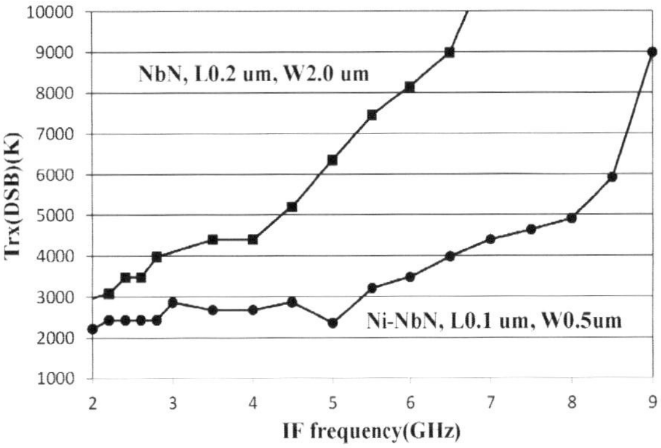

Fig. 1. Measured noise IF bandwidth of the NbN and the Ni-NbN HEBM at a LO frequency of 1.94 THz. The Ni-NbN HEBM shows the wider bandwidth of ~7 GHz.

References

[1] A. Kawakami, Y. Irimajiri, T. Yamashita, S. Ochiai, Y. Uzawa, "Broadening the IF band of a THz hot electron bolometer mixer by using a magnetic thin film," *IEEE Trans. THz Sci. Technol.*, vol.8, no.6, pp.647–653, Nov. 2018.

[2] A. Kawakami, Y. Irimajiri, "2 THz got electron bolometer using a magnetic thin film," *in Proc. 30th Int. Symp. Space THz Techn.*, Gothenburg, Sweden, Apr. 2019.

[3] Y. Irimajiri, A. Kawakami, "Measurements of Receiver Noise Temperature of an Ni-NbN HEBM at 2 THz," *in Proc. 30th Int. Symp. Space THz Techn.*, Gothenburg, Sweden, Apr. 2019.

Collaborative Research Laboratory of Terahertz Technology, Terahertz Technology Research Center, National Institute of Information and Communications, Koganei, Tokyo, Japan.

A Full Octave-Band OMT for Millimetre-Wave Receivers

Doug Henke[1]

There is a desire for an increased signal bandwidth within millimetre-wave receivers used in radio astronomy—beyond the single-mode range of rectangular waveguide. One approach is to combine more circuitry using MMICs, as wideband components can be designed using strip transmission lines. Ridged waveguides may also be used to extend the single-mode bandwidth.

In contrast to MMIC receivers, waveguide components are bulkier, but do allow for receiver architectures to be broken up into individual blocks rather than a fully integrated chip.

As a step towards a complete ridged waveguide receiver architecture (e.g., sideband-separating), a turnstile OMT using ridged waveguide is shown in Fig. 1, employing a T-junction combiner, E-bends, and a turnstile junction. The design assumes a platelet-type layout using 6 stacked pieces. Platelet assemblies are suitable for receiver array configurations and are compatible with 2.5-D manufacturing techniques spanning up to THz frequencies.

Here, it is assumed that the OMT will be directly milled and a minimum filleting radius was chosen as 0.8 mm (~0.133×free-space wavelength of the highest frequency). A cloverleaf shape results for the feed horn port, derived from a quad-ridge guide with filleted edges. Simulated performance is shown across 25–50 GHz in Fig. 2, and it is expected that the machined prototype (with transition) will demonstrate 15 dB return loss and 35 dB cross-polarisation separation.

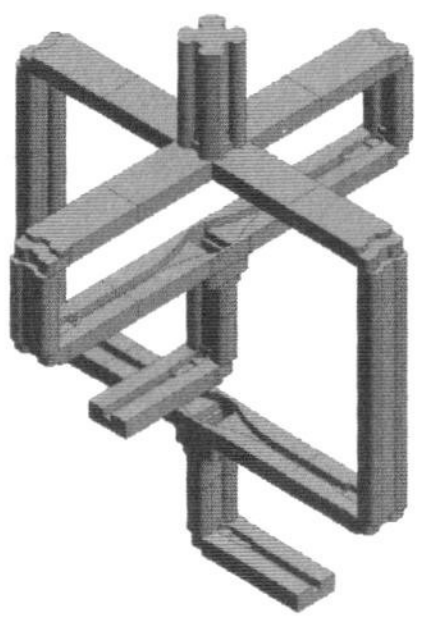

Fig. 1. CAD model of a turnstile OMT using ridged waveguide.

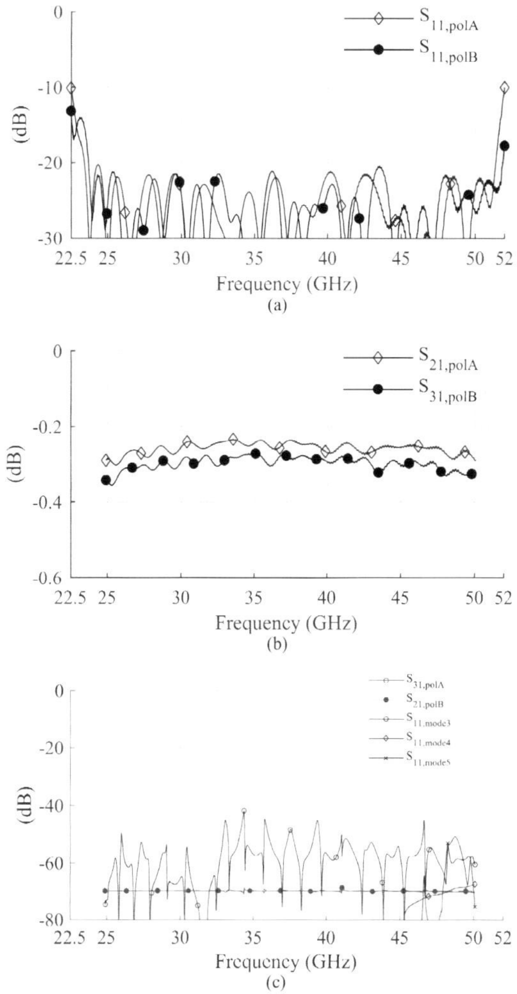

Fig. 2. Simulated performance of a ridged waveguide turnstile OMT: (a) reflected power; (b) insertion gain using a surface conductivity of 1.5E7 S/m; and (c) cross-pol and higher-order reflected modes.

[1] NRC Herzberg Astronomy and Astrophysics Research Centre, Victoria, BC, V9E 2E7, Canada.

NOTES:

A balloon-borne heterodyne receiver for atmospheric studies of atomic oxygen

M. Wienold[1], A. Semenov[1], H. Richter[1], H.-W. Hübers[1,2]

Atomic oxygen is the dominant species in the mesosphere and lower thermosphere (MLT) region of Earth's atmosphere. It plays an important role for the chemistry and energy balance of the MLT [1]. Its main coolant line is the $^3P_1$-$^3P_2$ ground state transition at 4.75 THz. The line shape of this transition cannot be resolved with typical Fourier-transform spectrometers. Due to water absorption, this line can also not be observed from ground. Currently, the only instrument which is capable of observing and resolving this transition is the upGREAT heterodyne receiver (German Receiver for astronomy at THz frequencies [2]) on board of SOFIA, the Stratospheric Observatory for Infrared Astronomy. However, observations are limited to night time and the radiometric accuracy is limited by the residual water vapor column still present at flight altitude.

We propose a balloon-borne heterodyne receiver, which is dedicated to the observation of atomic oxygen in the atmosphere. The instrument will allow for a higher radiometric accuracy than upGREAT/SOFIA, due to the much higher flight altitude of stratospheric balloons. It will also enable observations of diurnal variations. Figure 1(a) depicts simulated line profiles as expected for different elevation angles, and Fig. 1(b) shows the absorption profile as expected for column density measurements against the sun.

The instrument will be based on a dewar with a liquid helium stage for the receiver front end (hot-electron-bolometer (HEB) mixer and low-noise amplifier) and a solid nitrogen stage to cool the quantum-cascade laser used as local oscillator [3]. The Si lens of the quasi-optical HEB mixer will allow for an angular resolution of 30'. A pointing mirror will be used to measure at different elevation angles. The instrument will comprise a calibration unit. A filter unit will block most of the radiation from the sun, in order to allow for column density measurements against the sun. The backend of the receiver is a digital Fourier transform spectrometer.

The total weight of the instrument will be approximately 60 kg, which allows for combinations with other instruments on a typical 150-kg payload gondola as provided by the European HEMERA program.

Fig. 1. (a) Simulated OI emission line shape at night for different elevation angles as measured with a ballon-borne heterodyne spectrometer (b) OI line profile with a hot background (sun at sunrise). NRLMSISE-00 data for Nov. 20th 2018, Kiruna N67.8°E20.2°.

References

[1] M. G. Mlynczak and S. Solomon, "A detailed evaluation of the heating efficiency in the middle atmosphere," *J. Geophys. Res.*, vol 98, 10517–10541, 1993.

[2] C. Risacher, R. Güsten, J. Stutzki, H.-W. Hübers *et al.*, "The upGREAT Dual Frequency Heterodyne Arrays for SOFIA," *J. Astron. Instrum.*, vol 7, 1840014, 2018.

[3] H. Richter, M. Wienold, L. Schrottke, K. Biermann, H. T. Grahn, and H.-W. Hübers, "4.7-THz local oscillator for the GREAT Heterodyne Spectrometer on SOFIA," *IEEE Trans. Terahertz Sci. Technol.*, vol 5, 539–545, 2015.

[1] German Aerospace Center (DLR), Institute of Optical Sensor Systems, Rutherfordstr. 2, 12489 Berlin, Germany

[2] Humboldt-Universität zu Berlin, Department of Physics, Newtonstr. 15, 12489 Berlin, Germany

NOTES:

Research on High Precision Carbon Fiber Reinforced Plastics Reflector Panels for Dome A 5m Terahertz Explorer

Yuan Qian[1,2], Xufeng Hao[3], and Hongtao Xu[3], etc.

Dome A 5m Terahertz Explorer (DATE5) is a proposed telescope to be deployed at Dome A, Antarctica to explore the excellent terahertz observation condition unique to the site[1,2]. The low expansion property of carbon fiber reinforced plastics (CFRP) makes it more superior for the material of DATE5 antenna in such an extremely environment with large temperature difference.

In order to meet the high surface accuracy requirement of 10 microns rms for DATE5 primary reflector, there are two key challenges need to be settled in the development of high precision CFRP panel, which are the high precision molding accuracy and the high thermal stability respectively.

The research on high precision CFRP sandwiched panel has been developed by Purple Mountain Observation (PMO), CAS since 2014[3]. In order to achieving high replication efficiency from the mold to improve initial surface accuracy of panel, process of additional replication by resin layer is also applied [4]. Huge surface error such as fiber print though on surface, residual curvature error caused by difference of CTE between panel and mold appeared after the first molding process with high press and temperature. As a treatment, a layer of room-temperature curable resin with low curing shrinkage (2.5%) is slightly pressed in front of sheet after the first molding process. The surface of the rich resin coating with a thickness of 0.1mm~0.2mm is well adequate to eliminated most of the surface error, as shown in Fig.1.

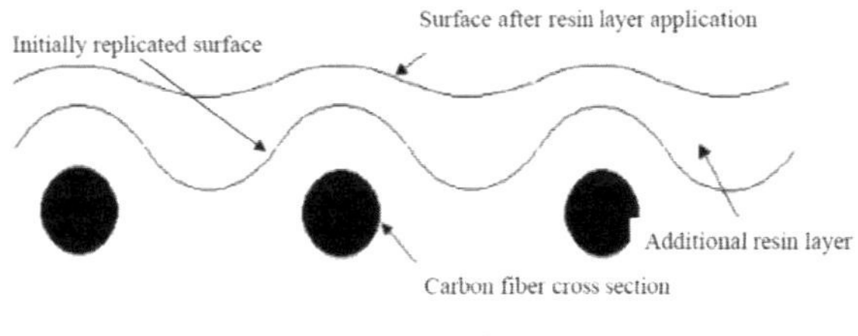

Fig.1 Primary reflector of microwave satellite

Meanwhile, A parametric finite element model (FEM) of DATE5 reflector panel is established base on the basic material properties of carbon fiber and resin used to predict its thermal deformation behavior at low temperature. The material properties are test at full temperature and the thermal deformation of the prototype panels are measured in a climate chamber. The thermal deformation behavior of the panel can be predicted more accurately by FEM after parameters corrected by experiments.

Three types of panels with different core structures have been studied by PMO in the past five years, which are CFRP sheets collocated with aluminum honeycomb core, CFRP tubes array core and CFRP lattice core respectively, as shown in Fig 2. The initial surface accuracy of prototype panels with 1 meter size can be kept within 10 microns rms stably now. The curvature of surface shape of panel with aluminum honeycomb core will change easily at low temperature caused by the thermal deformation in the normal direction of aluminum honeycomb. And the thermal stability of all-CFRP panel structures are much better, the thickness and fiber volume content of sheets are key performance parameters which affect the thermal stability.

(a) Aluminum honeycomb core

(b) CFRP tubes array core (c) CFRP lattice core

Fig.2 Three types of prototype panels

[1] Purple Mountain Observatory, Chinese Academy of Sciences, Nanjing 210033 China.

[2] Key Laboratory for Radio Astronomy, Chinese Academy of Sciences, Nanjing, 210033 China.

[3] Shanghai Composites Science & Technology Co.,Ltd, CASC, Shanghai, 201112 China.

REFERENCES

[1] Shi S C, Paine S, Yao Q J, et al. Terahertz and far-infrared windows opened at Dome A in Antarctica. *Nature Astronomy*. vol.1, no.1, pp. 1-7, 2016.

[2] Yang J, Zuo Y X, Zheng L, et al. Conceptual design studies of the 5m terahertz antenna for Dome A, Antarctica, *Research in Astronomy & Astrophysics*, vol.13, no.12, pp.1493-1508, 2013.

[3] Qian Y, Lou Z, Hao X F, et al. Initial development of high-accuracy CFRP panel for DATE5 antenna, *Proc. SPIE 9912, Advances in Optical and Mechanical Technologies for Telescopes and Instrumentation II, 99121D*, July 22, 2016.

[4] Massarello J J, Welsh J S, Hochhalter J D, et al. Fiber print-through mitigation technique for composite mirror replication. *Optical Engineering*, vol.45, no.12, pp.123401, 2006.

Suppressing cavity resonances in high-frequency amplifiers with metamaterial structures

D. Monasterio[1,2], N. Castro[3], F. Pizarro[3] and F. P. Mena[1].

Packaging active components (discrete transistors or microwave-monolithic circuits) into usable amplifiers is prone to produce resonances within the operation band, especially at high frequencies. In fact, the active components and its auxiliary circuitry have to be placed inside metallic cavities that, in principle, must be selected sufficiently small as not to allow the propagation of cavity modes within the frequency range of operation. However, as frequency increases, the dimensions of the required circuitry become, unavoidably, electrically large making difficult, if not impossible, to construct cavities that do not allow the propagation of the undesirable modes.

One of the most common methods to reduce those modes is to place an absorbing material above the circuitry. However, it is not efficient as the larger the gain of the amplifier is, the stronger the resonance. A more efficient method has appeared recently, the use of metamaterials [1]. The idea is to create an artificial magnetic wall with a periodic structure that generates a stop-band on the frequencies of the undesired cavity modes. The method has been used in amplifiers operating up to 50 GHz [2].

Recently, some of us presented a compact sideband separating downconverter appropriate for use in focal plane arrays [3]. The downconverter features a wideband mixer with an amplification stage in a novel architecture that allows obtaining an excellent return loss and good conversion gain in the W band. Despite of those excellent results, the mixer did not perform as intended as there was a clear reduction of the sideband rejection ratio between 84 and 88 GHz. Extensive simulations demonstrated that the problem originated in a resonance appearing in the cavity containing the amplification chip, a commercial monolithic circuit. Experimental results confirmed that there indeed existed a resonance in the cavity that did not disappear even with the use of a lid with an absorber. In this paper we will demonstrate that the use of a suitable artificial magnetic wall, instead of the absorber, suppresses the undesired resonances. Fig. 1 summarizes some of the results that will be presented at the conference.

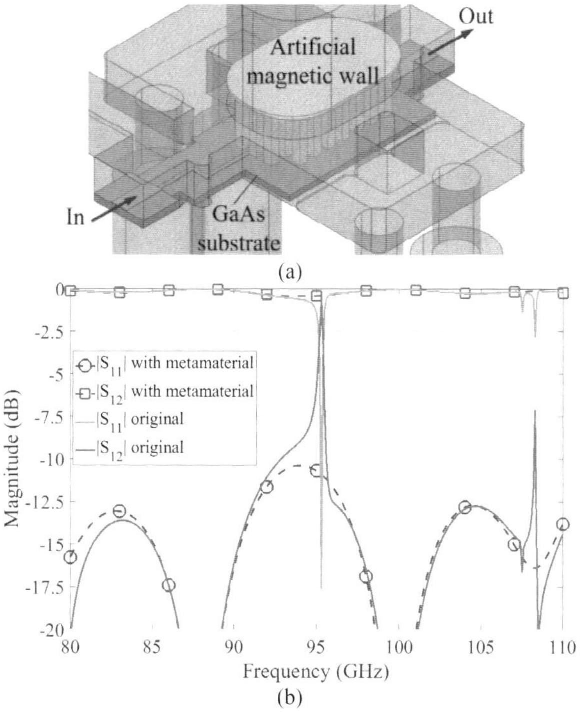

Fig. 1. (a) Drawing of the cavity containing the amplifier. The lid containing the artificial magnetic wall is placed above the amplifier and can be replaced by one containing an absorber. Both situations were simulated in HFSS. For simplification, only the substrate and a microstrip line connecting the input and output ports were drawn in the simulator. (b) Simulated transmission through the cavity when absorbing material (solid lines) and an artificial magnetic wall (dashed lines) are used to close the cavity.

[1] Electrical Engineering Department, Universidad de Chile, Av. Tupper 2007, Santiago, Chile.
[2] Astronomy Department, Universidad de Chile, Camino El Observatorio 1515, Santiago, Chile.
[3] Escuela de Ingeniería Eléctrica, Pontificia Universidad Católica de Valparaíso, Valparaíso, Chile.

References

[1] A. Rajo-Iglesias, A. U. Zaman & P.S. Kildal, "Parallel Plate Cavity Mode Suppression in Microstrip Circuit Packages Using a Lid of Nails," *IEEE Microw. Wirel. Co.*, vol. 20, no. 1, pp. 31–33, Jan. 2010.

[2] D. Henke, F. Jiang & S. Claude, "Mode Suppressing Packaging for 50 GHz Cryogenic Low-Noise Amplifiers," *Proc. 7th. European MICC*, pp. 623–626, Oct. 2012.

[3] D. Monasterio et al., "A Compact Sideband Separating Downconverter With Excellent Return Loss and Good Conversion Gain for the W Band," *IEEE Trans. THz Sci. Tech.*, vol. 9, no. 6, pp. 572–580, Nov. 2019.

Modelling of Travelling-Wave Kinetic-Inductance Parametric Amplifiers Implemented with Artificial Transmission Lines

F. P. Mena[1], and D. Valenzuela[1].

The use of the kinetic-inductance present in superconductors has bring about sensitive devices for use in diverse applications including radio astronomy. One of the most recent of such devices is the travelling-wave parametric (TKIP) amplifier. In this amplifier the superconducting material is fabricated (over a substrate) in the form of a transmission line that, owing to the kinetic inductance of the superconductor, has a nonlinear inductance. In such a medium a nonlinear parametric process can occur where several waves can exchange energy. Amplification results if only a limited number of waves is allowed to travel in the line. This is achieved by introducing adequate periodic perturbations in the line that produce stop bands at the selected frequencies.

Since it is relatively easy to fabricate, a CPW line was selected for the first implementation of a TKIP amplifier [1]. Despite of the promising results, one problem arose due to this particular selection of line. Given the high kinetic inductance needed to produce amplification, only lines with high impedance could be fabricated. Although efforts were made in matching the line, its connection to an external 50-Ω line resulted in large in the gain that render the device impractical.

One way to reduce the impedance, but maintaining the use of CPW lines, is to construct them in the form of artificial transmission lines [2]. In this kind of line, small sections of a CPW are used to recreate a transmission line made of lumped-elements. Here we will present, first, a model to construct such lines. The validity of the model was confirmed with the use of HFSS and Sonnet. We will also present designs of TKIP amplifiers using this lines. Table 1 and Fig. 1 show some of the results that will be presented at the conference.

Table 1. Two different designs of unperturbed lines.

	Design 1		Design 2	
	Impedance	$f_{\text{cut-off}}$	Impedance	$f_{\text{cut-off}}$
	Ω	GHz	Ω	GHz
Model	53.2	1287	36.3	898
Sonnet	58.8	---	43.0	---
HFSS	57.6	---	46.4	---

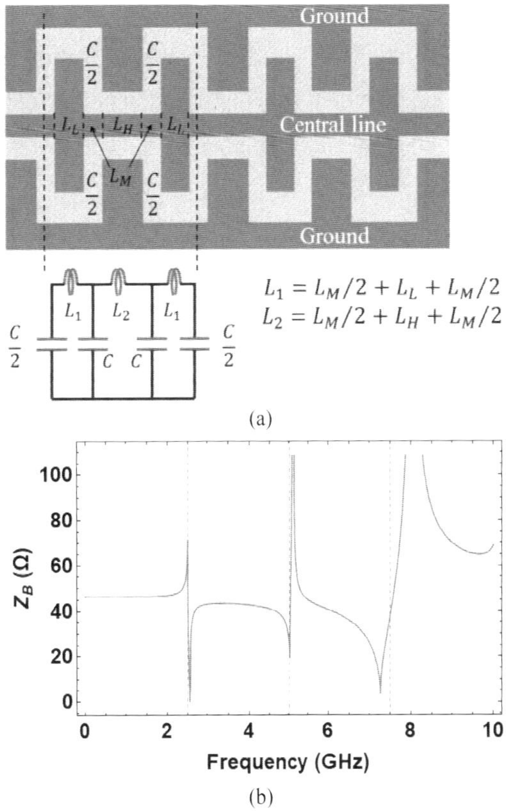

Fig. 1. (a) Circuital model of a unit cell of a (non-perturbed) superconducting artificial line made of CPW sections. (b) Bloch impedance of a perturbed line that can be used to implement a TKIP amplifier using four-wave mixing. Its dimensions were chosen as to obtain a Bloch impedance close to 50 Ω in the intended operational range (0–5 GHz).

REFERENCES

[1] B. H. Eom, P. K. Day, H. G. LeDuc & J. Zmuidzinas, "A wideband, low-noise superconducting amplifier with high dynamic range," *Nature Physics*, vol. 8, pp. 623–627, Aug. 2012.

[2] S. Chaudhuri et al., "Broadband parametric amplifiers based on nonlinear kinetic inductance artificial transmission lines," *Appl. Phys. Lett.*, pp. 152061, Apr. 2017.

[1] Electrical Engineering Department, Universidad de Chile, Av. Tupper 2007, Santiago, Chile.

This work was supported by Conicyt through grants Fondecyt 1180700 and Basal AFB-170002.

A Dispersion-Engineered Josephson Travelling Wave Parametric Amplifier with Periodic Impedance Perturbation

Kitti Ratter*† and Boon-Kok Tan*
*Department of Physics (Astrophysics), University of Oxford, Keble Road, Oxford OX1 3RH, UK
†Contact: kitti.ratter@physics.ox.ac.uk, phone +44 1865 273 303

***Abstract*—We present a new approach to develop a Josephson Junction Travelling Wave Parametric Amplifier (JTWPA) that could potentially minimise the gain-ripple effect. Our design consists of a 50 Ω superconducting niobium coplanar waveguide (CPW) periodically loaded with series of Josephson junctions (JJs) to provide the non-linearity required for wave mixing. The embedded JJs alter the characteristic impedance of the transmission line abruptly, therefore creating the stopbands in the transmission (S_{21}) for the suppression of higher harmonics and provide the means for dispersion engineering required to achieve exponential gain. The simulated gain profile of the amplifier shows that we can obtain a minimum of 15 dB gain from 4–12 GHz, close to a 100% bandwidth performance. More importantly, the characteristic impedance of the main linear transmission line remains unaffected by the strong pump, therefore ensuring the TWPA remains impedance matched to the input and output ports. This can reduce the unwanted gain undulation that inflict the optimal performance of the TWPA.**

I. Introduction

The development of cryogenic low-noise, broadband amplifiers is motivated by wide ranging technologies that require ultra-sensitive detection of weak microwave signals, such as radio astronomy [1], [2], [3] and quantum computing [4], [5], [6]. These applications generally use a high electron mobility transistor (HEMT) amplifier to amplify the weak microwave signals. However, the added noise of HEMT amplifiers is still 5–10 times the quantum limit, and their heat dissipation is high. In recent years, there has been an intense interest in developing an amplifier with quantum-limited noise performance that can achieve similar high gain over a broad bandwidth. The superconducting travelling wave parametric amplifier (TWPA) is an ideal candidate for realising such a device. TWPAs achieve high gain by wave mixing which transfer energy from a strong pump tone to the detected signal tone in a very long, ultra-low loss non-linear transmission line. The non-linearity is typically introduced by either the kinetic inductance of a superconducting wire (KITWPA), or by Josephson junctions embedded along the line. An alternative approach to realise a JTWPA is to use superconducting quantum interference devices (SQUIDs). This type of amplifier is generally referred to as SQUID-based TWPA (STWPA).

To achieve exponential gain, the power dependent phase difference between all the propagating tones needs to be minimised. This can be done by engineering the dispersion relation of the transmission line. In KITWPAs, the impedance can be periodically varied along the line to achieve phase matching, termed as the periodic loading scheme [7]. In JTWPAs, where the phase difference between the propagating waves is larger, resonators can be capacitively coupled to the transmission line [8], [9]. Dispersion engineering can also be achieved by the periodical modulation of the SQUID area in STWPAs [10]. However, this approach requires the careful control of the magnetic flux, which could complicate the experimental setup.

Generally, the non-linear inductance of a JJ is much larger than the kinetic inductance of a superconducting wire, hence JTWPAs and STWPAs require a shorter transmission line and lower pump power than KITWPAs for similar gain [8], [10]. This results in a better noise performance compared to KITWPAs, because even though the superconducting thin film displays only a minuscule surface resistivity, a high current passing through a very long line could still induce a non-negligible resistive loss in terms of I^2R.

Another challenge yet to overcome is matching the TWPA to a 50 Ω environment. Experimental gain profiles of TWPAs reported in the literature have been shown to suffer from gain ripples which arise from the reflections caused by the imperfect impedance matching at the input and output port of the amplifier [7], [8], [10].

Here, we present a new method to design a JTWPA that could potentially be better matched to a 50 Ω environment, while retaining its low noise performance. A unique feature of our TWPA design is that the main transmission of our device is made up of a linear passive circuit component in the form of a CPW, hence the characteristic impedance of the line will not be altered by the strong pump. This in turn ensures that the impedance between our device and the input/output 50 Ω interfaces is well-matched, minimising the gain-ripples caused by reflections. Furthermore, we use periodic loading instead of a large number of resonators to achieve phase matching, simplifying the design significantly. The loading sections, comprising of sets of JJs, are embedded along the centre conductor of the CPW at equal distances, locally altering the characteristic impedance, as well as serving as the non-linear medium for wave mixing. These periodic sections create stopbands in the transmission (S_{21}) and provide the required excursion in the otherwise linear dispersion relation. Eliminating the need for resonators not only greatly simplifies the fabrication process to increase yield, it also helps to reduce the footprint of the amplifier.

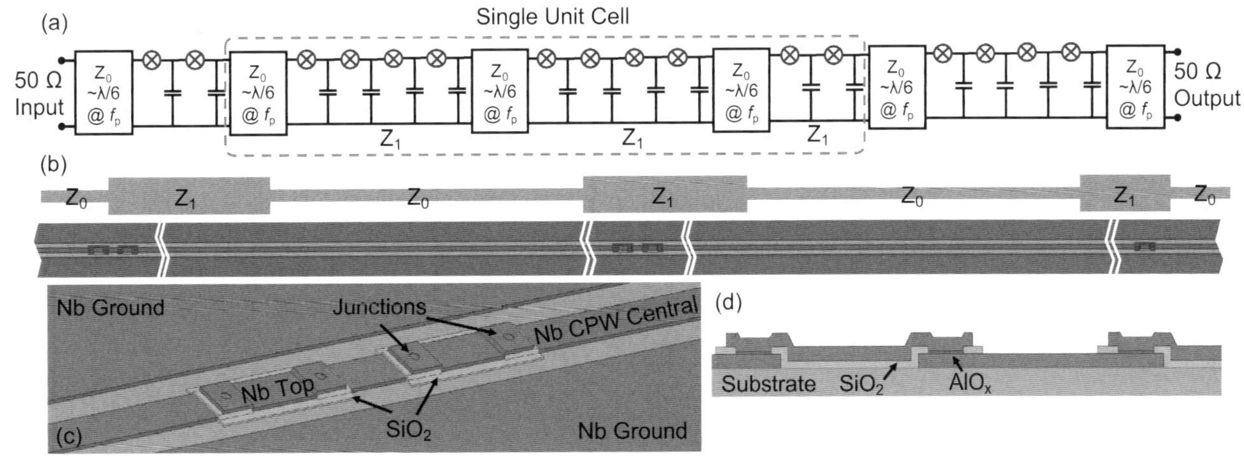

Fig. 1. (a) Circuit diagram: CPW is matched to the 50 Ω input/output. The unit cell consists of three CPW sections and three sets of JJs where one set has fewer (or more) junctions. The total length is half the wavelength of the central frequency. (b) Schematics of the amplifier: sets of JJs are periodically embedded in the centre conductor of a 50 Ω CPW, locally altering its characteristic impedance to Z_1 (c)&(d) Drawings show how the tunnel junction can be embedded along the CPW line. Gaps are created along the central conductor of the CPW, where thin layer of SiO_2 is deposited to protect the junction, before the top Nb layer is used to connect the CPW line.

II. Design Concept

There are two approaches to realise the periodic loading scheme with JJs. One option is to equally space the JJs along the transmission line, and periodically alter the junction size (thus the characteristic impedance) of a few of them to create the required stopbands and dispersion gap. However, this approach is hard to realise in reality. The transmission line is long compared to the junction size, thus the number of junctions required would be formidable from the fabrication point of view. To reduce the number of JJs, they could be distributed sparsely along the main transmission line, but in this case, they no longer behave as distributed elements, but rather as discrete lumped elements, and exponential parametric gain may not be possible.

A better solution is therefore to embed sets of JJs only at the loading sections. The junctions in a set are packed closely together to ensure that they behave as a distributed element. The spacing between the sets is chosen to be approximately one sixth of the wavelength (λ_c) of the central/pump frequency. This creates a large stopband at three times the pump frequency to suppress the unwanted third harmonic that may cause a shock wave in the amplifier. Additionally, every third of these sets have a reduced (or increased) number of junctions, to create the auxiliary stopbands and divergence in the wave vector near the pump frequency for exponential gain.

Fig. 1(a) and (b) illustrate the concept of our periodically loaded JTWPA. The low loss transmission line is expected to be realised as a niobium (Nb) CPW line, shown in green, in which the geometry will be optimised to obtain 50 Ω characteristic impedance. At the loading sections, junctions are embedded in the central conductor, altering the characteristic impedance from Z_0 to Z_1 at every $\sim \lambda_c/6$. The junction sections will be designed such that their dimensions match the width of the central conductor, as shown in Fig. 1(c). The single unit cell consists of three 50 Ω CPW sections, three sets of JJs where the third set has fewer (or more) junctions, as highlighted by the dashed box in Fig. 1(a). The length of such a unit cell corresponds to half the wavelength of the central frequency.

The tunnel junctions considered here comprise of niobium superconducting electrodes sandwiching an aluminium-oxide tunnelling barrier, as shown in Fig. 1(d). We expect to fabricate the junction array using the standard trilayer technology via Self-aligned Niobium Etch Process (SNEP) [11], [12], as shown in Fig 1(c) and (d). First, the Nb-AlO_x-Nb trilayer is formed across the entire wafer. Next, the trilayer is etched away to pattern the ground plane and the 'bottom' centre CPW strip line. The junctions are then defined by etching off the top Nb layer of this remaining trilayer apart from the area where the junctions are located. Before depositing the top Nb 'wiring' layer, a thin (few tens of nanometer) layer of SiO_2 is deposited around the junctions and in the gaps along the 'bottom' centre CPW strip line, to ensure that there is no electrical contact between the bottom and the top junction electrodes. Finally the 'top' Nb layer is deposited to complete the fabrication of the TWPA.

The choice of using Nb as the main superconducting material to form the transmission line and the JJs is mainly motivated by our aim to develop an ultra-low noise amplifier as readout for our superconductor-insulator-superconductor (SIS) receiver. This approach would allow the amplifier to operate at liquid helium bath temperatures, therefore a convenient integration is possible with SIS mixers. Higher gap temperature superconductor can be used as well in theory, but as mentioned earlier these type of high gap superconducting films will unavoidably have higher surface resistivity which could cause higher losses. Furthermore, fabrication of high-gap superconducting tunnel junction still remains a very challenging task [13], [14]. The use of Nb technology will also allow us to further integrate the TWPA with the SIS mixer on a single chip forming an integrated receiver in the future.

III. Simulation & Analysis

Based on the idea presented above, we have designed a periodically loaded JTWPA with a series of identical 1 μm^2 tunnel junctions embedded along a CPW line. The critical current of the junction is set to $I_C = 3.24\,\mu$A, with junction inductance $L_J = 101.48$ pH, and junction capacitance $C_J = 72.7$ fF. Each tunnel junction is shunted with a capacitor of 12 fF, increasing the impedance of the loading section from 50 Ω to about 93 Ω, creating a large and abrupt impedance perturbation. In each unit cell, the 50 Ω line is interrupted by two sets of 4-JJ arrays, and a third set of 2-JJ array loading sections. The distance between these sets of JJ arrays is approximately $0.15\lambda_c$ with the 4-JJ array occupying about $0.02\lambda_c$ and the 2-JJ array about $0.01\lambda_c$, resulting in a total of $\lambda_c/2$ long unit cell.

The combination of these design parameters produces deep stopbands in the S_{21} transmission profile and causes the dispersion near these gaps to diverge exponentially away from the otherwise linear relation, as shown in Fig 2(a). Here, we cascaded 70 unit cells together to form the JTWPA, which means a total of 700 junctions are required for each amplifier. The pump frequency is set to be near 8 GHz to obtain the extra dispersion required to compensate for the phase mismatch of all the waves travelling down the line. The large stopband near 24 GHz is created to suppress the third harmonic of the pump wave.

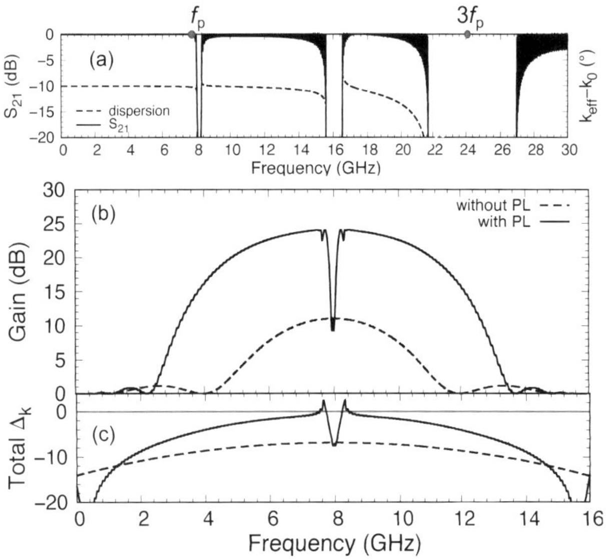

Fig. 2. (a) The dispersion relation of the amplifier shows divergencies where there is a stopband in the S_{21} profile. The ripple in the S_{21} curve is caused by the large mismatch between the loading section and the 50 Ω transmission line which is compounded due to the cascade of a large number of unit cells. (b) Gain profile simulated for the periodically loaded JTWPA comprising of 70 unit cells. It shows that at least 15 dB of gain is achievable across approximately 100% bandwidth. The pump frequency is set to 9.79 GHz, with a pump power of $I_p/I_C = 0.65$. The dashed curve shows the JTWPA without periodic loading scheme, but with the same number of JJs embedded uniformly along the CPW line. (c) The total phase mismatch between all waves i.e., the pump, signal, and idler, as the function of the frequency. Without the phase matching technique, the minimal phase difference is far from $\Delta_k = 0$, hence high exponential gain is not possible. By setting the $f_p = 9.79$ GHz, we obtain the additional dispersion correction required to achieve $\Delta_k \rightarrow 0$ across a broad bandwidth, hence achieving high gain over wide operational bandwidth.

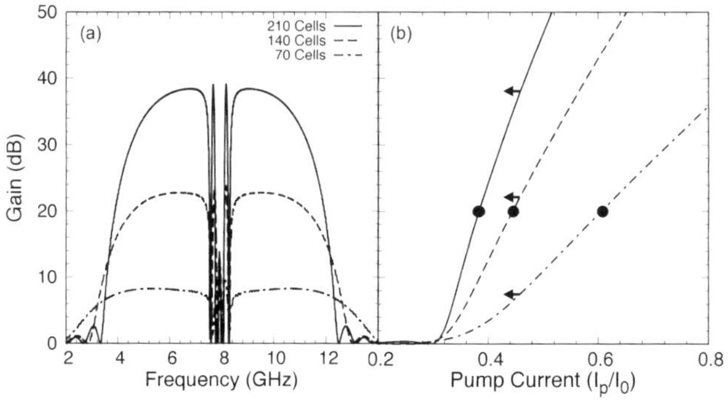

Fig. 3. (a) The gain-bandwidth product of periodically loaded JTWPAs with three different lengths: 70, 140 ,and 210 unit cells. The pump current is set to $I_p = 0.46I_C$ for all curves simulated. As expected, with longer transmission length, the interaction time between the signal and the pump waves is longer, hence higher gain is achievable. (b) The parametric gain as the function of the pump current for the three different cases. The left-pointing arrows show the gain at 7 GHz, where $I_p = 0.46I_C$. To achieve 20dB gain at 7 GHz, the 210-cell design requires only $I_p = 0.4I_C$, whereas the 140-cell design requires $I_p = 0.48I_C$, and the 70-cell case requires $I_p = 0.64I_C$.

Fig. 2(b) shows the gain curve of our periodically loaded JTWPA. As can be observed, we managed to achieve more than 15 dB gain from 4–12 GHz, a 3:1 (100%) gain-bandwidth product, with the maximum gain approaching 25 dB. This high gain is made possible by setting the pump frequency near the first stopband, around 8 GHz, where the dispersion increases rapidly. By carefully choosing the position of the pump frequency (f_p=9.79 GHz), we can reduce the total phase mismatch to be close to zero ($\Delta_k \rightarrow 0$) across a large bandwidth, as shown in Fig. 2(c). The corresponding pump current required in this case is approximately $I_p = 0.65I_C$. For comparison, we plot the gain-bandwidth product and the phase mismatch of the same JTWPA without the periodic loading scheme, i.e. uniformly spreading the same number of JJs along the 50 Ω line, while keeping the same pump wave settings. It is clear that without the periodic loading scheme, the mismatch Δ_k is too large for high gain at all frequencies, therefore not only the maximum gain is reduced from close to 25 dB to about 10 dB, but the relative bandwidth is reduced significantly as well.

As mentioned, high pump current is undesirable since it might incur high resistive loss in the form of I^2R, thus degrading the noise performance of the amplifier. For relatively thin Nb film at microwave frequency, the surface resistance should be minuscule at cryogenic temperature, compared to a high gap superconducting film of the same thickness. However, depending on the quality of the superconducting Nb film, the surface resistive loss could still be significant, especially in the case of a TWPA where the transmission length is long. This is particularly concerning for operation at liquid helium bath temperature, at only 40% of the critical temperature of Nb. A potential solution to eradicate this predicament is to reduce the pump power at the expense of a longer transmission length. This means the number of Josephson junctions needs to be increased as well. Fig. 3(a) shows the comparison of the achievable gain for three different periodically loaded JTWPAs with different lengths: the original design comprising of 70 unit cells, a 140-cell design, and a 210-cell design (700 JJs, 1400 JJs, and 2100 JJs, respectively). In this example, we fixed the pump power to be half of the value previously considered, i.e., $I_p/I_C = 0.46$, and adjusted the pump frequency for optimal gain. As shown in Fig. 3(a), with longer transmission length, we can achieve a higher gain, as expected. In the case of the 140-cell design, the gain-bandwidth product is almost the same as for the original design, but requiring only half the pump power. This is further illustrated in Fig. 3(b) where the left-pointing arrows show the peak gains of 8, 22, and 38 dB for the 70-cell, the 140-cell, and the 210-cell designs, respectively. Another way to depict the effect of longer transmission lengths is shown by the black dots in Fig. 3(b). To achieve 20 dB peak gain, a 210-cell JTWPA requires only about $I_p/I_C = 0.36$ while the 70-cell design requires close to $I_p/I_C = 0.6$, 2.78× higher pump power. Nevertheless, it is worthwhile noting that the longer transmission length may complicate the fabrication of such a TWPA, as the number of tunnel junctions required becomes much higher which might affect the fabrication yield. However, with the mature Nb tunnel junction technology, we believe that TWPAs with such high number of tunnel junctions are realisable.

IV. Conclusion

We have presented a broadband Josephson junction travelling wave parametric amplifier with close to 15 dB-100% gain-bandwidth product. Phase matching is achieved by periodically embedding sets of JJs in the centre conductor of a linear 50 Ω coplanar waveguide. Because of the linearity of the main transmission medium, this amplifier has a great promise to display a ripple-free experimental gain curve. Although the design presented in this paper was focused on using Nb as the main superconducting material for both the transmission line and the tunnel junction, the simplicity of the design means that it can be adapted for other superconducting materials as well. For example, replacing niobium with aluminium would make the amplifier suitable for quantum bit (qubit) platforms that are being developed for quantum computation and information applications. These platforms generally use aluminium technology to construct the qubits and the corresponding superconducting circuitry, therefore on-chip integration of an aluminium TWPA is feasible without complicating the fabrication procedure.

References

[1] C. Jia, L. Zhiqun, L. Qin, and W. Zhigong, "A 1–16 GHz wideband CMOS low-noise IF amplifier for ALMA radio astronomy receivers," in *2011 IEEE 13th International Conference on Communication Technology*, pp. 1066–1069, IEEE, 2011.

[2] D. M. Smith, L. Bakker, R. H. Witvers, B. E. Woestenburg, and K. D. Palmer, "Low noise amplifier for radio astronomy," *International Journal of Microwave and Wireless Technologies*, vol. 5, no. 4, pp. 453–461, 2013.

[3] C. Bockstiegel, J. Gao, M. Vissers, M. Sandberg, S. Chaudhuri, A. Sanders, L. Vale, K. Irwin, and D. Pappas, "Development of a Broadband NbTiN Traveling Wave Parametric Amplifier for MKID Readout," *Journal of Low Temperature Physics*, vol. 176, pp. 1–7, 08 2014.

[4] R. Vijay, D. H. Slichter, and I. Siddiqi, "Observation of Quantum Jumps in a Superconducting Artificial Atom," *Phys. Rev. Lett.*, vol. 106, p. 110502, Mar 2011.

[5] M. H. Devoret and R. J. Schoelkopf, "Superconducting Circuits for Quantum Information: An Outlook," *Science*, vol. 339, no. 6124, pp. 1169–1174, 2013.

[6] M. Esposito, J. Rahamim, A. Patterson, M. Mergenthaler, J. Wills, G. Campanaro, T. Tsunoda, P. Spring, S. Sosnina, S. Jebari, K. Ratter, G. Tancredi, B. Vlastakis, and P. Leek, "Development and characterization of a flux-pumped lumped element Josephson parametric amplifier," *EPJ Web Conf.*, vol. 198, p. 00008, 2019.

[7] B. H. Eom, P. K. Day, H. G. LeDuc, and J. Zmuidzinas, "A wideband, low-noise superconducting amplifier with high dynamic range," *Nature Physics*, vol. 8, no. 8, pp. 623–627, 2012.

[8] C. Macklin, K. O'Brien, D. Hover, M. E. Schwartz, V. Bolkhovsky, X. Zhang, W. D. Oliver, and I. Siddiqi, "A near–quantum-limited Josephson traveling-wave parametric amplifier," *Science*, vol. 350, no. 6258, pp. 307–310, 2015.

[9] T. C. White, J. Y. Mutus, I.-C. Hoi, R. Barends, B. Campbell, Y. Chen, Z. Chen, B. Chiaro, A. Dunsworth, E. Jeffrey, J. Kelly, A. Megrant, C. Neill, P. J. J. O'Malley, P. Roushan, D. Sank, A. Vainsencher, J. Wenner, S. Chaudhuri, J. Gao, and J. M. Martinis, "Traveling wave parametric amplifier with Josephson junctions using minimal resonator phase matching," *Applied Physics Letters*, vol. 106, no. 24, p. 242601, 2015.

[10] L. Planat, A. Ranadive, R. Dassonneville, J. Puertas Martínez, S. Léger, C. Naud, O. Buisson, W. Hasch-Guichard, D. M. Basko, and N. Roch, "Photonic-Crystal Josephson Traveling-Wave Parametric Amplifier," *Phys. Rev. X*, vol. 10, p. 021021, Apr 2020.

[11] P. Pütz, *Fabrication of superconductor-insulator-superconductor devices for heterodyne mixer applications with electron beam lithography*. PhD thesis, Universität zu Köln, 2004.

[12] P. K. Grimes, *Design and analysis of 700 GHz finline mixers*. PhD thesis, University of Cambridge, 2006.

[13] A. Karpov, D. Miller, F. Rice, J. Stern, B. Bumble, H. LeDuc, and J. Zmuidzinas, "Low Noise 1 THz–1.4 THz Mixers Using Nb/Al-AlN/NbTiN SIS Junctions," *IEEE Transactions on Applied Superconductivity*, vol. 17, no. 2, pp. 343–346, 2007.

[14] A. Khudchenko, A. M. Baryshev, K. I. Rudakov, P. M. Dmitriev, R. Hesper, L. de Jong, and V. P. Koshelets, "High-gap Nb-AlN-NbN SIS junctions for frequency band 790–950 GHz," *IEEE Transactions on Terahertz Science and Technology*, vol. 6, no. 1, pp. 127–132, 2015.

High Reflectance of Roughened Surface for the Integrating Sphere of SAFARI Calibration Source

Ming-Jye Wang[1], Chun-Lun Wang[1], Chuang-Ping Chiu[1], Ting-Hang Pei[1]

The SPICA mission (SPace Infrared telescope for Cosmology and Astrophysics) is a proposed satellite equipped an infrared space observatory with a large, cryogenically cooled telescope which will provide infrared spectroscopy with the wavelength range of mid- (12~36 μm, SMI) and far-infrared (34~250 μm, SAFARI) and far-infrared polarimetry in 3 bands (110 μm, 220 μm, 350 μm, B-BOP). SAFARI (SpicA FAr-infraRed Instrument) is an imaging Fourier transform spectrometer with a 2'×2' instantaneous field of view for 34~250 um. It will offer the sensitivity of the R ~300 (SAFARI/LR mode) about 5×10^{-20} Wm^{-2} (5σ/1h) assuming a TES (Transition Edge Senor) detector NEP (Noise Equivalent Power) of 2×10^{-19} W/√Hz [1]. A calibration source assembly is coupled to SAFARI to perform absolute calibration. The output signal of calibration source is geometrically diluted by an integrating sphere (IS). Our design of calibration source assembly consists of two microlamps and IS. The output signal of IS is expected to be a blackbody radiation and be homogeneously illuminated. The inner surface of IS must be rough to randomly scatter the THz light in the sphere. The loss of roughened surface has to be as low as possible to prevent the degradation of output power of IS. Here we will present the progress of IS for SAFARI calibration source assembly.

We have created rough surface on the inner side of aluminum semi-sphere with various methods. To increase the reflectance, we deposited Au film of 1 μm after surface is roughened. We measured the output spectrum of 5cmϕ IS at room temperature in FTS (Fourier transform spectroscopy) with commercial InSb cryogenic detector made by QMC Instrument Ltd. Similar to SRON's early result [2], the loss of sandblasted surface is high. We attribute such loss to the highly twisted surface after bombardment. The estimated reflectance is only about 0.63 if the reflectance of smooth surface is 0.96. This issue was solved by adding the aluminum wet etching process after sandblasting. Figure 1 compares the output spectra of IS with smooth (red line) and roughed (black line) surfaces in the wavelength range of 30-300 μm. The IS with roughened surface has a slightly low output intensity at short wavelength region (< 100μm). The ratio of their intensity is about 1.4. The estimated reflectance of roughened surface is 0.93 if the value of smooth surface is 0.96. In long wavelength region, the output intensity difference between two ISs is negligible. Our result indicates the loss of is significantly reduced if the sandblasted surface is treated by aluminum wet etching process additionally. The spatial uniformity of IS output power was measured. The measured pattern agrees the calculated pattern based on blackbody radiation assumption. The detail will be present in this paper.

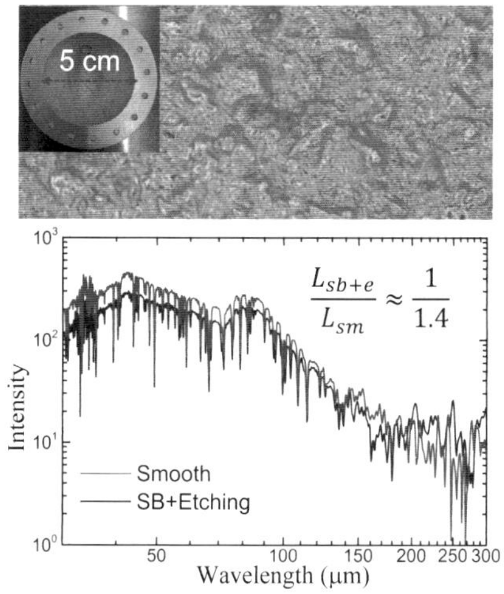

Fig. 1. Upper: The photo of roughened surface of integrating sphere. Lower: The comparison between the output spectra of integrating spheres with smooth (red) and roughed (black) surfaces. Their intensity ratio is about 1.4 at wavelength shorter than 100 μm and close to 1 at long wavelength region. The estimated surface reflectance of roughed surface is about 0.93 if the value of smooth surface is 0.96.

References

[1] P. Roelfsema, H.Shibai, L. Armus, D. Arrazola, M. Audard, M. Audley, ... I. Yamamura, (2018). "SPICA—A Large Cryogenic Infrared Space Telescope: Unveiling the Obscured Universe". *Publications of the Astronomical Society of Australia*, 35, E030, 2018, doi:10.1017/pasa.2018.15

[2] W.M. Laauwen, C. de Jonge, L. Ferrari, and P. Dieleman, "A Calibration Source for SAFARI on-ground Calibration", https://asd.gsfc.nasa.gov/conferences/FIR/posters/Roelfsema_SAFARI-Calibration.pdf, May 2014

[1]Institute of Astronomy and Astrophysics, Academia Sinica, Taipei, 10617, Taiwan.

NOTES:

Mid-infrared multi-beam local oscillator source based on a fiber coupled quantum cascade laser

Yuan Ren[1,2], DaiXi Zhang[1,2,3], Zheng Wang[1,2], KangMin Zhou[1,2], JiaQiang Zhong[1,2], Dong Liu[1,2], Wei Miao[1,2], Wen Zhang[1,2], ShengCai Shi[1,2]

High resolution spectroscopy has been proven to be a powerful tool in atmospheric and astronomical research, ranging from millimeter wavelength to mid-infrared wavelength range. As demonstrated at millimeter/sub-millimeter wavelength region, multi-beam heterodyne array receivers play a vital role, not only due to its improved mapping speed but also with the enhanced sensitivity in the continuum observation mode. However, at mid-infrared wavelengths, multi-beam heterodyne array receiver has not been reported yet, which is largely due to the complexity of efficiently demultiplexing the local oscillator (LO) source.

We have developed a fiber coupled demultiplexed local oscillator source for cryogenic application based on a distributed feedback quantum cascade laser (QCL) at an operating wavelength of 10.6 μm. Reflection phase grating generating 2×2 beams has been achieved with a total power efficiency of 64% based on gold covered etched silicon pattern. The diffraction beam has been readily coupled into a 2 meter long polycrystalline fiber with a core diameter of 240 μm, with 49% coupling efficiency achieved using an aspheric lens. Despite the 48% transmission coefficient for the polycrystalline fiber, Gaussian-like beam with a FWHM of 200 μm has been delivered at the fiber output. By aligning the fiber core position to the center of a superconducting hot electron bolometer mixer placed at 4 K in a closed-cycle cryostat, we demonstrate the application of fiber coupled demultiplexed QCL source for pumping the superconducting mixer with adequate LO power.

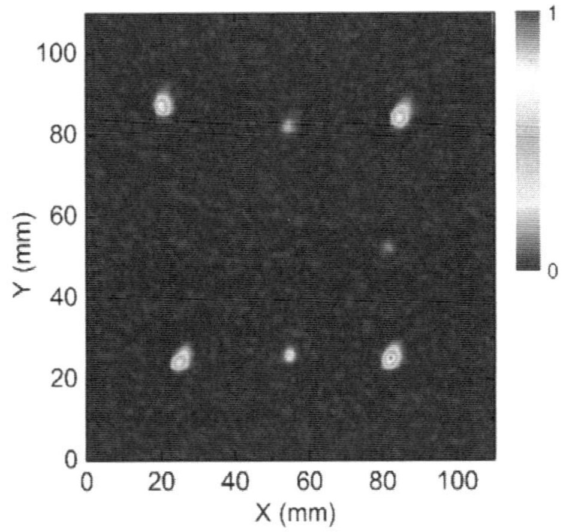

Fig. 1. Experimental result of diffraction pattern at 10.6 μm for the step phase grating.

[1] Purple Mountain Observatory, Chinese Academy of Sciences, 10th YuanHua Road, Nanjing, 210033, China

[2] Key Lab of Radio Astronomy, Chinese Academy of Sciences, 10th YuanHua Road, Nanjing, 210033, China

[3] University of Science and Technology of China, 96th Jinzhai Road, Hefei, 230026, China

Characterization System for SIS Frequency Converters based on Scalar Mixer Calibration Technique

Takafumi Kojima[1], Yoshinori Uzawa[1], and Wenlei Shan[1], Yuto Kozuki[2]

It is well known that a quasiparticle superconductor-insulator-superconductor (SIS) tunnel junction potentially allows positive conversion gain in the down-conversion process from millimeter wave to microwave. In recent, it was found that the tunnel junction could also have positive up-conversion gain from microwave to millimeter-wave [1]. Utilizing this bilateral gain in the SIS up- and down-conversion processes, we demonstrated a microwave amplification with low power consumption in several micro-watts using a proof-of-concept test module. The module uses SIS up- and down-converters which were connected in cascade and driven by an identical local oscillator (LO) power. The superconducting microwave amplifier based on this concept is promising for future radio astronomical heterodyne instruments such as a large scale multi-beam receiver at (sub-)millimeter wavelengths [3], but it has not been studied well so far.

In order to design the amplifier with low noise and high gain, it is important to characterize both the SIS converters in radio frequency (RF) and intermediate frequency (IF). However, in general, the evaluation of cryogenic devices based on a module approach has been carried out at millimeter-wave and even microwave frequencies. In this case, the measured conversion gain and impedance are affected by interconnections and interfaces of the package, e.g. connectors, bonding wires and transmission line. This requires correction of the measured values to extract device parameters, which causes uncertainties.

We have established a characterization system for an SIS up- and down-converter using a 4-K cryogenic probe station in combination with a vector network analyzer up to 50 GHz (Fig.1). The characterization system allows us to directly measure mixer conversion gain and reflection coefficients at RF and IF. It consists of a high pass filter with f_{cutoff} of 26 GHz and a low pass filter with f_{cutoff} of 21.5 GHz to divide the RF and IF signals, and a broadband 20-dB coupler for LO power injection. The LO chain uses a microwave synthesizer and a frequency doubler within an output frequency of 26.5-40 GHz. A DC bias voltage to the SIS junction is applied via a bias tee. A vector network analyzer (N5225B PNA, 10 MHz-50 GHz) produced by Keysight technology was used to measure the frequency conversion properties.

For accurate characterization of conversion properties, a scaler mixer calibration technique was employed. An SIS device sample based on Nb/Al/AlO$_x$/Nb junctions with a diameter of 0.8 μm and critical current density of approximately 5.5 kA/cm^2 was used for the verification of the measurement technique.

In the conference, the characterization method and measurement example of the SIS converter will be presented.

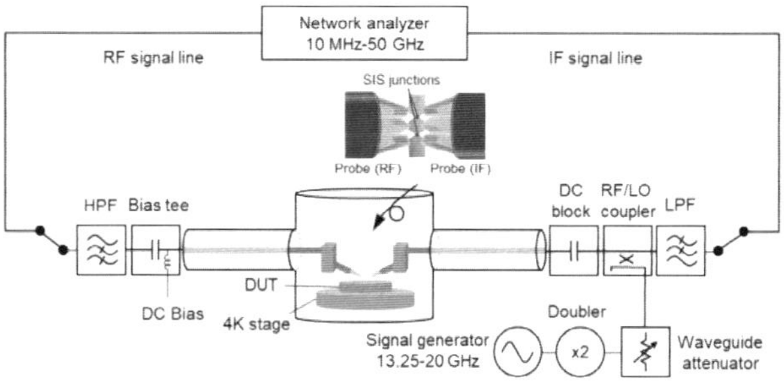

Fig. 1. Schematic of the on-wafer characterization system for an up- and down-converter.

ACKNOWLEDGEMENT

This work was supported by JSPS KAKENHI under Grant JP19H02205 and Grant JP18H03881.

REFERENCES

[1] Y. Uzawa, T. Kojima, W. Shan, A. Gonzalez, M. Kroug, "Investigation of SIS Up-Converters for Use in Multi-pixel Receivers," Journal of Low Temperature Physics, Vol. 193, pp 512–517, Nov. 2018.

[2] T. Kojima, Y. Uzawa, W. Shan, "Microwave amplification based on quasiparticle SIS up and down frequency converters," AIP Advances 8, 025206, 2018.

[3] W. Shan, S. Ezaki, K. Kaneko, A. Miyachi, T. Kojima and Y. Uzawa, "Experimental Study of a Planar-Integrated Dual-Polarization Balanced SIS Mixer," in IEEE Transactions on Terahertz Science and Technology, vol. 9, no. 6, pp. 549-556, Nov. 2019.

[1] National Astronomical Observatory of Japan, Mitaka, Tokyo, Japan.
[2] The University of Elector-communications, Chofu, Tokyo, Japan

Planar silicon metamaterial lenses with integrated anti-reflection coatings for frequencies around 150 GHz

J. Jauk[1,*], G. Wang[1,*], V. Moreno[1], M. Parioleau[1], A.L. Fontana[1], S. Leclercq[1], and E.F.C. Driessen[1]

For quasi-optical elements in the millimeter and sub-millimeter range, silicon is an interesting material. Its high refractive index facilitates the production of compact and lightweight elements. Moreover, its thermal conductivity allows better thermalization at cryogenic temperatures, and the loss tangent of bulk high-resistivity silicon (tan $\delta <$ 10-4) is without competition.

Silicon is however very difficult to machine, and the high refractive index necessitates the use of anti-reflection coatings. Micromachined anti-reflection coatings have been developed for planar substrates but become increasingly more difficult for curved surfaces of e.g. lenses.

In this work, we follow a different approach. We use the fact that it is possible to modulate the refractive index of a material by inserting sub-wavelength voids and changing the fill factor of the voids. This way, a silicon metamaterial with a dielectric constant between 3.3 and 11.7 can be generated [1].

From (visible) optics it is well known that a curved surface, such as a lens, can be mimicked by a planar element that has the appropriate refractive index gradient. Thus, we designed a planar silicon element, with an expected focal length of 180 mm. The lens has a diameter of 50 mm, and the effective dielectric constant varies continuously from 11.2 in the center to 3.3 at the edges of the lens, by the means of an hexagonal array of holes with a period of 104 μm and a hole size varying from 13.4 μm in the center to 81 μm at the edges. The total thickness of the lens is 1 mm, which is about an order of magnitude thinner than an equivalent (curved) polyethylene lens with a similar focal length.

The element was fabricated out of four 250 μm thick, high-resistivity silicon wafers that were micromachined using a Bosch process in an inductively coupled plasma etcher. The wafers were aligned using dowell pins, and pressed together in a dedicated holder. The same process was used to fabricate two anti-reflection coating layers, using 250 μm wafers with an adapted dielectric constant profile, such that the dielectric constant is given by $\varepsilon AR(x) = \sqrt{\varepsilon lens(x)}$. The thickness of the AR coating is chosen such that the averaged reflection over the surface of the lens is minimized, since the optimal quarter-wave adaptation is impossible using a planar design with varying dielectric constant. The modular design of our devices easily allows for more intricate AR coatings consisting of multiple layers of varying thickness and dielectric constant.

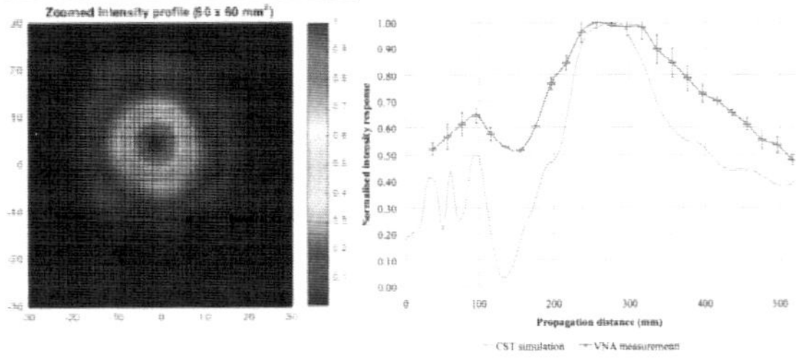

Fig. 1. (left) Measured beam profile of a microwave feed horn imaged by the silicon metamaterial lens. (right) On-axis intensity as a function of distance, measured (red) and as simulated with CST (green). Both measurements and simulations are made at a frrequency of 150 GHz.

The fabricated lens was subsequently characterized in an antenna range. We find that the imaging properties of the lens are excellent, but that the effective focal length is approximately 30% smaller than designed. The measured value is confirmed by CST simulations of the structure (see Fig. 1). This difference is attributed to the fact that the lens is of comparable thickness to the wavelength, privileging near-field effects, and rendering ray-optical approaches such as used for the design insufficient. Simple transmission measurements indicate the effectiveness of the AR coating.

The presented technology offers great perspective in terms of compact, planar, low-loss optics. Moreover, the technology can be easily integrated with silicon detector wafers, and future developments that involve more elaborate anti-reflection coatings, integrated filtering, or microlens arrays, are just part of the possibilities.

We will discuss further optimization of this technology towards actual optical elements for radio-astronomy applications, as well as the current limitations of the used technology.

References

1. K.F. Schuster et al., "Micro-machined quasi-optical elements for THz applications", *16th International Symposium of Space Terahertz Technology*, p. 524-528 (2005).

[1] Institut de Radioastronomie Millimétrique, Grenoble, France
[*] These authors contributed equally to this work.

NOTES:

Development of wideband circular polarizer in 70-118 GHz band

Sho Masui[1], Yutaka Hasegawa[2], Toshikazu Onishi[1], Hideo Ogawa[1], Satoshi Ochiai[3], Issei Watanabe[3]

VLBI (Very Long Baseline Interferometer) observations generally require circular polarized detection. Since these radio signals are extremely weak, radio-astronomical receivers are usually operated at cryogenic temperatures to have very low noise performance. Septum type polarizers or a combination of an OMT (Orthogonal mode Transducer) and a 90-deg differential PS (Phase Shifter) can be employed in millimeter-wave receivers for dual-circular polarization observations. Septum type polarizer is usually used to observe circular polarization in satellite communications or low frequency VLBI receivers, because of its low insertion loss and compact structure. However, the fractional bandwidth of septum type polarizer is limited less than 25 %. On the other hand, a combination of an OMT and a 90-deg differential phase shifter has a potential for realizing much wider-bandwidth circular polarizer. Since OMTs have wideband performance generally [1], it is important to develop a broadband 90-deg phase shifter for wideband circular polarization observations in VLBI.

A reported fractional bandwidth of wideband PS is so far up to 30 % [2]. The fractional bandwidth of PS is limited by the frequency dependence of the phase shift. To decrease the frequency dependence of the phases, we examined whether a wideband phase shifter could be developed by combining PSs with different phase shift characteristics. Then we have discovered that ridge type PS and corrugated type PS have different characteristics. We have designed and fabricated wider band PS by combining the corrugated and ridge structures, and obtained a good performance with the fractional bandwidth of ~50% at 70-118 GHz (Fig. 1). We also designed and fabricated a wideband OMT with the fractional bandwidth of ~50% at 67-116 GHz. We are currently assembling to its wide band PS and OMT, and measuring the characteristics of wideband circular polarizer.

At the conference, we will present the measurement performance of wideband circular polarizer.

[1] Osaka Prefecture University, Sakai, 599-8531, Japan.
[2] Institute of Space and Astronautical Science, Sagamihara, 252-5210, Japan.
[3] National Institute of Information and Communications Technology, Koganei, 184-8795, Japan.

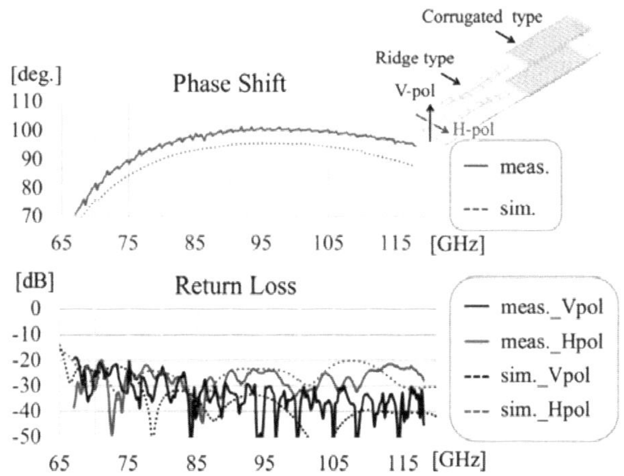

Fig. 1. Simulation and measurement performances of the broadband PS composed of ridge type and corrugated type PS. Upper graph shows the phase differences between each polarization. Lower graph shows the return loss.

REFERENCES

[1] Alvaro Gonzalez, "Double-Ridged Waveguide Orthomode Transducer (OMT) for the 67–116-GHz Band," *J Infrared Milli Terahz Waves*, vol. 39, pp 723–737, Aug. 2018

[2] Moon-Hee Chung, "Development of 85-115 GHz 90-deg Phase Shifter using Corrugated Square Waveguide," *Proceedings of the 44th European Microwave Conference*, pp. 1146-1149, Oct. 2014

NOTES:

Preliminary Design of a Multibeam Receiver for the SMA

John D. Garrett, Paul K. Grimes, and Cheuk-Yu Edward Tong

Abstract—**In these proceedings, we present the preliminary design of a multibeam receiver for the Submillimeter Array (SMA). The design is centered at 345 GHz, which was chosen to coincide with the scientifically important CO(3–2) emission line and its common isotopes. The design presented here has 3 pixels, but we plan to expand the array to 7 pixels in a hexagonal pattern in a future iteration. The array uses sideband separating (2SB) receivers, which greatly improves the mapping speed for narrow emission lines, when compared to equivalent double sideband receivers. Compared to the wSMA's planned high-band dual-polarization receiver, a 7-pixel 2SB multibeam system should provide $\sim$7 times faster mapping speed. This would allow for large-scale mapping of molecular emission lines in the Milky Way and nearby galaxies with very high angular resolution.**

Index Terms—**multibeam receivers, focal plane arrays, SIS mixers, superconducting detectors, submillimeter-wave astronomy**

I. Introduction

The Submillimeter Array (SMA) is an eight-element radio interferometer located near the summit of Mauna Kea, Hawaii, that operates from 180-420 GHz. It is currently undergoing a large-scale upgrade to improve the noise performance and increase the instantaneous bandwidth of its receivers [1], [2], [3]. Through the wideband SMA (wSMA) upgrade, the SMA antennas will be refitted with new cryostats housing new receiver cartridges. These receivers will operate over two bands: a low-band from 210–270 GHz and a high-band from 280–360 GHz. Among other benefits, the new receivers will improve the SMA's sensitivity to continuum sources and increase the number of emission lines that can be observed simultaneously. Since the new wSMA cryostat will be smaller than the current SMA cryostat, there will soon be enough room to host guest receivers inside the receiver cabins, allowing for the ability to test new receiver designs with the wSMA, including new multibeam systems.

Multibeam receivers (also called focal plane arrays) have the potential to increase the mapping speed of a telescope by a factor equal to the number of pixels N within the array (when compared to an equivalent single-beam design). However, the mapping speed is also inversely proportional to the system noise temperature T_{sys} squared:

$$\text{mapping speed} \propto \frac{N}{T_{\mathrm{sys}}^2} \tag{1}$$

John D. Garrett, Paul K. Grimes and Cheuk-Yu Edward Tong are with the Center for Astrophysics | Harvard & Smithsonian, Cambridge, MA 02138, USA (e-mail: john.garrett@cfa.harvard.edu; pgrimes@cfa.harvard.edu; etong@cfa.harvard.edu).

Therefore, any increase in the system noise temperature can quickly negate the speed improvement afforded by the array. This means that any multibeam system must add additional pixels without compromising on the state-of-the-art noise performance of modern single-beam systems. Additional challenges include efficiently coupling each pixel to the main dish, managing the added mechanical/electrical complexity, dividing and injecting the local-oscillator (LO) signal without interfering with the astronomical signal path, and processing the additional intermediate frequency (IF) signals. (Comprehensive discussions of focal plane arrays can found in [4], [5], [6].)

Overall, a multibeam receiver on the wSMA would allow for wide-field mapping of molecular emission lines at high angular resolution. The improved mapping capability would be beneficial for a variety of observations, including the study of star formation in the Milky Way and nearby galaxies. It would also be the first multibeam system on an interferometer, taking advantage of the wSMA correlator system. As such, it would be able to act as a pathfinder for future multibeam systems on other large submillimeter-wave interferometers, such as the Atacama Large Millimeter/Submillimeter Array (ALMA) and the Northern Extended Millimeter Array (NOEMA). Due to its modest size and the ability to host guest receivers, the wSMA provides an excellent environment to test and develop new multibeam receivers for submillimeter wave interferometers.

In this work, we discuss the requirements for a multibeam receiver on the wSMA in Sec. II, investigate the potential mapping speed improvement in Sec. III, and present a preliminary multibeam system in Sec. IV. Sec. V concludes this study.

II. Design Considerations

There are several design considerations, specific to the wSMA, which influence the optimal multibeam receiver architecture. First, the maximum size of an array on the wSMA is limited by the size of the Nasymth tubes within the antennas. A low-band array ($\nu \sim 230$ GHz) would be limited to $\sim$3 pixels in a triangular pattern and a high-band array ($\nu \sim 345$ GHz) would be limited to 7 pixels in a hexagonal pattern. Since both bands target important molecular emission lines (e.g., the CO(2–1) and the CO(3–2) transitions, respectively) and both have good atmospheric transmission at the SMA site, a 7-pixel array centered at 345 GHz presents a greater potential mapping speed improvement.

Secondly, the array will use a similar backend to the wSMA. Assuming a total processed IF bandwidth of 64 GHz, this limits the IF bandwidth to $\sim$4.6 GHz per pixel for a 7-pixel

2SB receiver or ~9.2 GHz per pixel for a 7-pixel DSB array. Since the total processed IF bandwidth will be identical to the wSMA's single-beam receivers, the primary advantage of the array will be mapping narrow emission lines (the continuum mapping speed will be similar to the wSMA). Therefore, image sideband rejection would help to lower the system noise temperature and further increase the mapping speed. This could be accomplished by using a sideband separating (2SB) mixing scheme, or by terminating the image sideband in a reactive or cryogenic load (e.g., using a narrow bandpass filter to only pass the desired signal).

III. Mapping Speed Improvement

In this section, we compare the mapping speed of different multibeam receiver architectures: double sideband (DSB) receivers and single sideband (SSB) receivers. Based on Eqn. 1, the mapping speed of two different multibeam receivers can be compared using:

$$\Gamma_{2,1} = \frac{N_2}{N_1}\left(\frac{T_{\rm sys,1}}{T_{\rm sys,2}}\right)^2 \tag{2}$$

where $T_{\rm sys}$ is the SSB system noise temperature[1] (including receiver and atmospheric noise) and N is the number of pixels in the array ($N = 1$ for a single-beam receiver).

For a ground based DSB receiver, the SSB system noise temperature is [7], [8]:

$$T_{\rm sys}^{\rm DSB} = \frac{2\left(T_{\rm rx}^{\rm DSB} + T_{\rm sky}\right)}{e^{-\tau}} \tag{3}$$

where $T_{\rm rx}^{\rm DSB}$ is the DSB receiver noise temperature[2], $T_{\rm sky}$ is the atmospheric brightness temperature, and $e^{-\tau}$ is the atmospheric transmission. Likewise, for a SSB receiver, the SSB system noise temperature is:

$$T_{\rm sys}^{\rm SSB} = \frac{2\,T_{\rm rx}^{\rm DSB} + T_{\rm sky}\left(1 + 1/R_{\rm sb}\right)}{e^{-\tau}} \tag{4}$$

where $R_{\rm sb} = G_s/G_i$ is the sideband ratio, i.e., the ratio of the signal sideband gain G_s to the image sideband gain G_i. For an ideal SSB receiver, $R_{\rm sb} \to \infty$ and Eqn. 4 reduces to:

$$T_{\rm sys}^{\rm SSB} = \frac{2\,T_{\rm rx}^{\rm DSB} + T_{\rm sky}}{e^{-\tau}} \tag{5}$$

Note that Eqns. 3–5 ignore several noise contributions, including the cosmic microwave background and spillover effects. The main beam efficiency and the loss of the sideband separation scheme are also ignored. More detailed $T_{\rm sys}$ calculations can be found in [8], [9].

Using Eqns. 3–5, the system noise temperature is plotted in Fig. 1 for SSB and DSB receivers. For this plot, we assumed $T_{\rm rx}^{\rm DSB} = 55$ K, which is the target value for the wSMA's single-beam high-band receiver [1], and we calculated $e^{-\tau}$ and $T_{\rm sky}$ using the *am* atmospheric model [10] with a precipitable water vapor of 1.86 mm (the median value at the SMA site).

[1]In this report, we always use the SSB system noise temperature, unless specified otherwise. The SSB $T_{\rm sys}$ is the most important figure-of-merit for observing narrow emission lines, which are only present in one sideband.

[2]$T_{\rm rx}^{\rm DSB}$ is the noise temperature value that is normally measured in the lab via the Y-factor technique (since the test signal is present in both sidebands.)

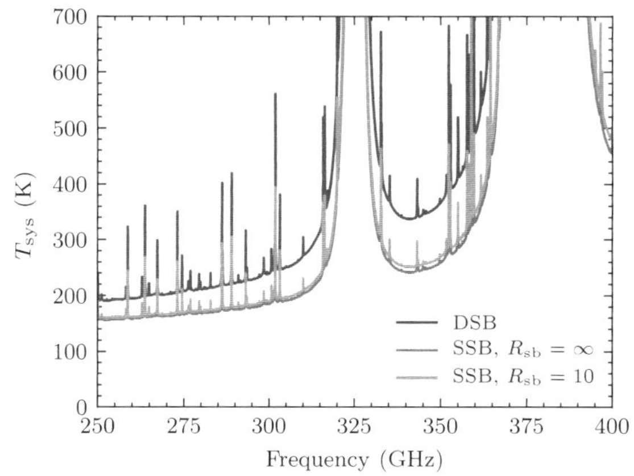

Fig. 1. Estimated SSB system noise temperature ($T_{\rm sys}$) of DSB and SSB receivers (Eqns. 3–5). For this plot, we assumed $T_{\rm rx}^{\rm DSB}$=55 K and a precipitable water vapor of 1.86 mm, which is the median value at the SMA site. The atmospheric properties were calculated using the *am* atmospheric model [10].

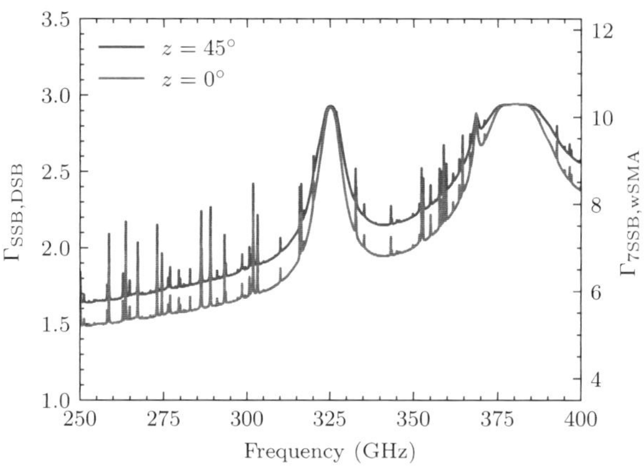

Fig. 2. Mapping speed ratio of a SSB receiver compared to a DSB receiver (left y-axis; calculated using Eqn. 2). This is plotted for two different zenith angles: z=0° and z=45°. The right y-axis compares a 7 pixel SSB multibeam receiver to the wSMA's single-beam DSB receiver.

As we can see, the ideal SSB receiver ($R_{\rm sb} \to \infty$) has a significantly lower $T_{\rm sys}$ value than the DSB receiver because it is able to remove the atmospheric noise contribution from the image sideband. This reduces the overall $T_{\rm sys}$ value by ~29% at 345 GHz. Even with a modest sideband ratio of $R_{\rm sb} = 10$, the atmospheric noise in the image sideband is reduced by 90% and $T_{\rm sys}$ is reduced by ~26%.

Based on Eqn. 2, lower system noise temperatures result in faster mapping speeds. In Fig. 2, the mapping speed of a SSB receiver is compared to a DSB receiver. As we can see, at 345 GHz, the SSB receiver has a mapping speed that is approximately twice as fast as the DSB receiver. Furthermore, the SSB receiver provides an even greater mapping speed improvement as the atmospheric transmission worsens (e.g., when there is a higher precipitable water vapor or a larger zenith angle).

The downside is that SSB receivers are much more complicated than DSB receivers. For example, image rejection can be accomplished by a tuneable waveguide backshort (e.g., HERA [11]), a Mach-Zehnder interferometer (e.g., HARP [12]), a Martin-Puplett interferometer (e.g., CHAMP+ [13]) or a cryogenic waveguide bandpass filter. These image rejection schemes generally have limited IF bandwidth and are very cumbersome. Another option is to separate the signal and image sidebands using a 2SB receiver. This has the added advantage of recovering both the upper and lower sidebands; however, it essentially doubles the complexity of the receiver. For example, a 2SB receiver requires a 90° hybrid for the RF signal, a power divider for the LO signal, two RF/LO diplexers and two SIS mixers per pixel. For our purposes, the added complexity of the 2SB receiver architecture is acceptable because our array will be a relatively modest 7 pixels.

We can rewrite Eqn. 2 to calculate the mapping speed improvement of a 7-pixel 2SB mutlibeam receiver when compared to the wSMA single-beam receiver[3]:

$$\Gamma_{7\mathrm{SSB,wSMA}} = \frac{7}{2}\left(\frac{T_{\mathrm{sys}}^{\mathrm{DSB}}}{T_{\mathrm{sys}}^{\mathrm{SSB}}}\right)^2 \tag{6}$$

This value is plotted in Fig. 2 using the y-axis on the right-hand side. As we can see, a 7-pixel 2SB multibeam receiver should be able to map the CO(3–2) emission line at 345.8 GHz approximately 7 times faster than the wSMA single-beam receiver.

IV. Preliminary Design of the Multibeam Receiver

The preliminary design of the multibeam receiver is shown in Figs. 3 and 4. The construction is similar to the array proposed for the Greenland Telescope [14], except that this new design has been expanded from DSB receivers to 2SB receivers. The design uses a combination of split-block waveguides (with the split in the E-plane of the waveguide) and stacked waveguide plates (where the direction of propagation is normal to the split plane). The design shown in Figs. 3 and 4 only has 3 pixels, but we plan to expand this array to 7 pixels in the near future, with the pixels arranged in a hexagonal pattern.

The RF (astronomical) signals enter the array block through the feed horns shown at the top of Fig. 4. RF quadrature hybrids then divide each RF signal into two separate waveguides with a 90° relative phase shift, which is necessary for sideband separation[4]. The LO signal enters from the side of the mixer block (Fig. 5). It is evenly divided into six waveguides using cascaded E-plane power splitters (0° relative phase shift), and then each LO waveguide is diverted 90° using H-plane waveguide bends to be parallel to the RF waveguides. Note that there are two LO waveguides per pixel. The RF and LO signals are then diplexed using silicon-mounted cross-guide couplers [15], and each RF+LO signal is coupled to an SIS

[3]Note that the wSMA single-beam receivers are dual-polarization. Assuming that polarization data is not important, $N_{\mathrm{wSMA}} = 2$.

[4]Alternatively, it is possible to divide the RF signals using in-phase power splitters and divide the LO signals using 90° RF hybrids. This can allow for wider RF bandwidth, but in-phase power splitters generally suffer from poor isolation and poor return loss, which in turn degrades R_{sb} and T_{sys}.

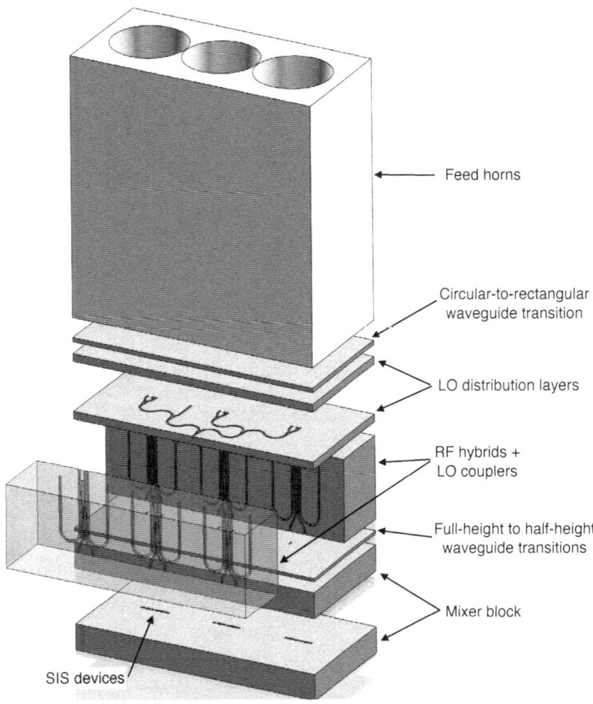

Fig. 3. Exploded view of the preliminary 3-pixel 2SB multibeam receiver. Most of the waveguide components are made from stacked waveguide plates, where the direction of propagation is normal to the split plane. The RF hybrids and LO direction couplers are made from a single split-block waveguide, where the split is in the E-plane of the waveguide.

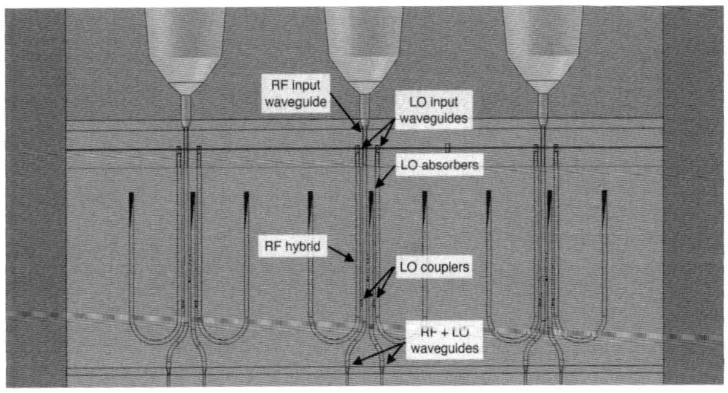

Fig. 4. Cross-section of the 3-pixel 2SB multibeam receiver. Each RF signal first passes through an RF hybrid, which divides the RF signal in two with a 90° relative phase difference. The LO signals are diplexed with the RF signals using silicon-mounted cross-guide couplers (the same coupler that is used by the wSMA's single-beam receivers [15]).

mixer device. Finally, the two down-converted IF signals from each pixel are combined using another 90° hybrid to recover the upper and lower sidebands (not shown). It is also possible implement this step in the digital backend of the telescope.

This multibeam receiver will be designed to cover the atmospheric window centered at ~345 GHz (seen in Figs. 1 and 2). Since most of the important molecular emission lines are situated between 329 GHz and 355 GHz, the multibeam receiver can get away with a narrower RF bandwidth than the wSMA's high-band receiver, without compromising on the utility of the receiver. For example, the CO(3–2), ^{13}CO(3–2)

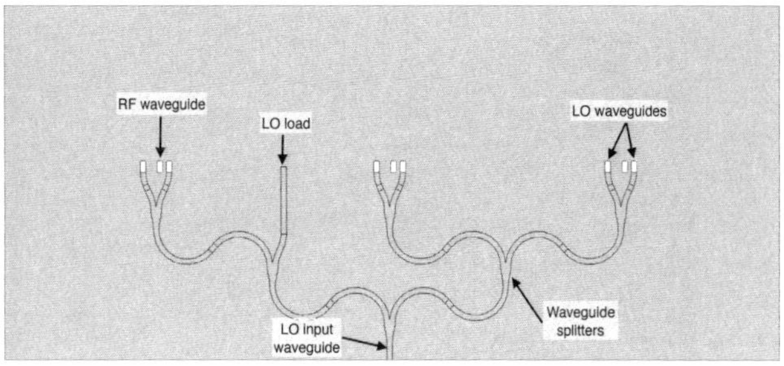

Fig. 5. Cross-section of the local-oscillator (LO) distribution layer. The LO signal enters from the bottom of this figure and then it is divided in six using cascaded E-plane power dividers. The LO waveguides are then diverted downwards using H-plane waveguide bends to be in parallel with the RF waveguides. There are two LO waveguides per RF waveguides. Note that the LO power dividers require excellent isolation to minimize cross-talk between the RF waveguides.

and $C^{18}O(3–2)$ transitions occur at 345.8 GHz, 330.6 GHz and 329.3 GHz, respectively. Furthermore, since the target objects have to have relatively large angular diameters for the array to be beneficial, the array will only be used on low-redshift objects ($z \sim 0$), meaning that we do not need to consider redshifted emission lines. Therefore, we can optimize the performance of the array from ~329–356 GHz to achieve the best possible noise temperature.

To achieve a high $R_{\rm sb}$ value, the array block will require careful fabrication, especially for the RF hybrids and the LO power dividers. Any machining errors could result in amplitude and phase imbalances that will lower the $R_{\rm sb}$ value and thereby cause a higher noise temperature.

Finally, note that these are preliminary drawings, meaning that many important features are missing from the CAD drawings in Figs. 3–5, e.g., coaxial cables, magnetic coils, etc. We are also considering several design modifications including using silicon-mounted RF hybrids to facilitate fabrication, using permanent magnets to suppress the Josephson currents and reduce the number of DC wires, and using integrated SiGe low noise amplifiers (LNAs) to minimize the heat load of the array.

V. Conclusion

We have presented the preliminary design of a multibeam receiver for the SMA. Once the wSMA upgrade is complete, this receiver could be mounted in the guest cryostat inside the wSMA receiver cabin. The proposed 7-pixel sideband separating receiver has the potential to increase the mapping speed of the wSMA by a factor of ~7. This would allow for large-scale surveys with the wSMA of molecular emission lines in the Milky Way and nearby galaxies.

References

[1] P. Grimes, E. Tong, S. Paine, L. Zeng, and R. Blundell, "Introducing the wSMA: A Major Upgrade to the SMA Receiver Systems," *Submillimeter Array Newsletter*, vol. 23, pp. 17–19, Jan. 2017.

[2] R. Rao, P. Grimes, E. Tong, and D. Wilner, "Science with the Upgraded ultra-wideband Submillimeter Array (wSMA) in the Next Decade," in *2019 URSI Asia-Pacific Radio Science Conference (AP-RASC)*, New Delhi, India, Mar. 2019.

[3] P. Grimes, R. Blundell, S. Paine, E. Tong, and L. Zeng, "Receivers for the wideband Submillimeter Array," in *Proceedings of the 31st International Symposium on Space Terahertz Technology (ISSTT)*, Tempe, AZ, Mar. 2020.

[4] C. E. Groppi and J. H. Kawamura, "Coherent Detector Arrays for Terahertz Astrophysics Applications," *IEEE Transactions on Terahertz Science and Technology*, vol. 1, no. 1, pp. 85–96, Sep. 2011.

[5] P. F. Goldsmith, "Sub-Millimeter Heterodyne Focal-Plane Arrays for High-Resolution Astronomical Spectroscopy," *The Radio Science Bulletin*, vol. 2017, no. 362, pp. 53–73, Sep. 2017.

[6] J. D. Garrett, "A 230 GHz Focal Plane Array Using a Wide IF Bandwidth SIS Receiver," DPhil dissertation, University of Oxford, 2018.

[7] C. Groppi, A. Baryshev, U. Graf, M. Wiedner, P. Klaassen, and T. Mroczkowski, "First Generation Heterodyne Instrumentation Concepts for the Atacama Large Aperture Submillimeter Telescope," in *Proceedings of the 30th International Symposium on Space Terahertz Technology (ISSTT)*, Gothenburg, Sweden, Apr. 2019.

[8] J. W. Lamb, "SSB vs . DSB for Submillimeter Receivers," *ALMA Memo 301*, Apr. 2000.

[9] A. Remijan, A. Biggs, P. Cortes, B. Dent, J. Di Francesco, E. Fomalont, A. Hales, S. Kameno, B. Mason, N. Philips, K. Saini, F. Stoehr, B. Vila Vilaro, and E. Villard, "ALMA Technical Handbook, ALMA Doc. 8.3, ver. 1.0," *ALMA partnership*, 2020.

[10] S. Paine, "The *am* atmospheric model," 2019. [Online]. Available: https://zenodo.org/record/3406483

[11] K.-F. Schuster, C. Boucher, W. Brunswig, M. Carter, J.-Y. Chenu, B. Foullieux, A. Greve, D. John, B. Lazareff, S. Navarro, A. Perrigouard, J.-L. Pollet, A. Sievers, C. Thum, and H. Wiesemeyer, "A 230 GHz heterodyne receiver array for the IRAM 30 m telescope," *Astronomy & Astrophysics*, vol. 423, no. 3, pp. 1171–1177, Sep. 2004.

[12] J. V. Buckle, R. E. Hills, H. Smith, W. R. F. Dent, G. Bell, E. I. Curtis, R. Dace, H. Gibson, S. F. Graves, J. Leech, J. S. Richer, R. Williamson, S. Withington, G. Yassin, R. Bennett, P. Hastings, I. Laidlaw, J. F. Lightfoot, T. Burgess, P. E. Dewdney, G. Hovey, a. G. Willis, R. Redman, B. Wooff, D. S. Berry, B. Cavanagh, G. R. Davis, J. Dempsey, P. Friberg, T. Jenness, R. Kackley, N. P. Rees, R. Tilanus, C. Walther, W. Zwart, T. M. Klapwijk, M. Kroug, and T. Zijlstra, "HARP/ACSIS: a submillimetre spectral imaging system on the James Clerk Maxwell Telescope," *Monthly Notices of the Royal Astronomical Society*, vol. 399, no. 2, pp. 1026–1043, Oct. 2009.

[13] C. Kasemann, S. Heyminck, A. Bell, A. Belloche, C. Castenholz, R. Gusten, H. Hafok, A. Henseler, S. Hochgurtel, T. Klein, I. Kramer, A. Korn, K. Meyer, D. Muders, F. Pacek, F. Schafer, G. Schneider, G. Wieching, H.-J. Wunsch, A. Baryshev, R. Hesper, T. Zijlstra, C. Lodewijk, and T. Klapwijk, "CHAMP+: A Powerful Submm Heterodyne Array," in *Proceedings of the 19th International Symposium on Space Terahterz Technology (ISSTT)*, Groningen, Netherlands, Apr. 2008, pp. 166–172.

[14] P. K. Grimes, K. Asada, R. Blundell, R. Burgos, H.-H. Chang, M. T. Chen, D. Goldie, C. Groppi, C. C. Han, P. T. P. Ho, Y. D. Huang, M. Inoue, D. Kubo, P. Koch, J. Leech, E. de Lera Acedo, P. Martin-Cocher, H. Nishioka, M. Nakamura, S. Matsushita, S. N. Paine, N. Patel, P. Raffin, W. Snow, T. K. Sridharan, R. Srinivasan, C. N. Thomas, C.-Y. E. Tong, M.-J. Wang, C. Wheeler, S. Withington, G. Yassin, and L.-Z. Zeng, "Instrumentation for single-dish observations with The Greenland Telescope," in *SPIE Astronomical Telescopes + Instrumentation*, vol. 9153, Jul. 2014, p. 91531V.

[15] L. Zeng, E. Tong, and P. Grimes, "Waveguide Components for wSMA Frontends," in *Proceedings of the 31st International Symposium on Space Terahertz Technology (ISSTT)*, Tempe, AZ, Mar. 2020.

Performance of the SIS terahertz photon detector

Hajime Ezawa[1,2], Hiroshi Matsuo[1,2], Masahiro Ukibe[3], Go Fujii[3], and Shigetomo Shiki[3]

We are developing a fast terahertz photon detector. Fast photon detector may be one of the key technologies for future terahertz instruments, which may resolve each photons to realize high sensitivity. In addition, fast photon detectors may be used to study the photon statistics, which may act as a precise measure to determine the physical properties of the radiation.

Among various technologies to detect terahertz radiation, SIS junction (or STJ) is a good candidate as a device to realize a fast photon detector, which should be as fast as 1 GHz. We have developed an SIS junction of Nb/Al/AlOx/Al/Nb which exhibits a leakage current as low as 1 pA at a cryogenic temperature of T <0.7 K [1]. The junction was 3 μm x 3 μm in size, with critical current density of J_c=300 A/cm^2. The junction is developed using the CRAVITY facility at AIST [2]. The junction has been integrated into an antenna coupled SIS detector. The design of the detector is discussed in [1]: A twin-slot antenna connects to the SIS through a coplanar waveguide. The SIS forms a parallel connected twin junction (PCTJ) [3], which is tuned to our target frequency of 500 GHz. The low critical current density will result in a relatively narrow RF bandwidth (4 GHz for our case). The current detector is aimed for laboratory experiments where the background radiation dominates, therefore the narrow bandwidth is preferred.

Recently, we evaluated the performance of the detector. The leakage current was measured at bias voltage of 600 μV to confirm the low leakage feature of 1-2 pA. The optical performance was also evaluated: The SIS detector was irradiated with a blackbody source to evaluate the photocurrent. The measurement was done at cryogenic temperature of 0.4 K, with the SIS detector mounted on a silicon hyper-hemispherical lens. The spectral response was investigated with a Fourier Transform Spectrometer. The results clearly show that the detector is sensitive to terahertz photons, however the performances were slightly shifted from the designed values. The optical efficiency was derived to be low, which may have caused by the degraded spectral performance, or transmission properties due to the detector design.

[1] National Astronomical Observatory of Japan, Tokyo 181-8588, Japan.
[2] Department of Astronomical Science, The Graduate University for Advanced Studies, SOKENDAI, Tokyo 181-8588, Japan.
[3] National Institute of Advanced Industrial Science and Technology, Tsukuba, Ibaraki 385-8568, Japan.

Based on the results, we are now in the course to revise the detector design: The detector fabrication parameters are investigated to realize a better spectral performance. The geometrical design of the detector is studied to improve the transmission efficiency from the antenna to the detector, and to realize a stable fast readout. The performances of the detectors with the revised design are being studied, which includes the optical response of the detectors with improved fabrication parameters. The design and performance of the SIS photon detectors, and their improvement studies will be discussed in the presentation.

References

[1] H. Ezawa, H. Matsuo, M. Ukibe, G. Fujii, S. Shiki, "Studies on Terahertz Photon Counting Detectors with Low-Leakage SIS Junctions", J. Low Temp. Phys. 194, 426-432, 2019, DOI: 10.1007/s10909-019-02149-y.

[2] M. Ukibe, S. Shiki, Y. Kitajima, M. Ohkubo, "Soft X-ray Detection Performance of Superconducting Tunnel junction Arrays with Asymmetric Tunnel Junction Layer Structure", Jpn. J. Appl. Phys. 51, 010115, 2012, DOI: 10.1143/JJAP.51.010115.

[3] T. Noguchi, S-C. Shi, J. Inatani, "Parallel Connected Twin SIS Junctions for Millimeter and Submillimeter Wave Mixers: Analysis and Experimental Verification", IEICE Trans. Electron. E78-C, 481-489, 1995.

Waveguide Components for wSMA Frontends

Lingzhen Zeng[1], C. Edward Tong[1], and Paul K. Grimes[1]

***Abstract*—We report on the development of the next generation of wideband Submillimeter Array (wSMA) low band and high band frontend waveguide components: from the design, fabrication, to the measurement of the prototypes. We also report on the ongoing effort of the integration of these frontend components into a receiver cartridge.**

***Index Terms*—Directional couplers, frontend, orthomode transducers, waveguide twists**

I. Introduction

The Submillimeter Array (SMA) is an eight antenna interferometer array on the summit of Mauna Kea, Hawaii. The array is undergoing a major upgrade to become the wideband Submillimeter Array (wSMA). The wSMA upgrade will include a new cryostat with two receiver cartridges for each antenna: the low band receiver will be operated by a local oscillator (LO) tunable between 210 GHz and 270 GHz; while the LO for the high band receiver will be tunable between 280 GHz to 360 GHz. Each receiver will offer a wideband intermediate frequency (IF) of 4 GHz to 20 GHz [1].

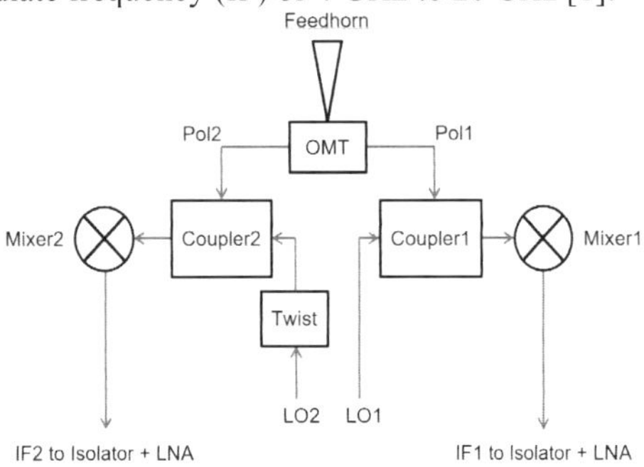

Fig. 1 Schematic diagram of wSMA frontends

As shown in figure 1, the frontend of each wSMA receiver consists of a number of waveguide components, including a profiled-corrugated feedhorn, a Bøifot type orthomode transducer (OMT), two silicon-chip-based LO couplers and a rectilinear waveguide twist. The LO couplers couple the LO signal in to the sky signal coming from the OMT outputs. The waveguide twist rotates the E-plane of the LO2 port for compact integration. The OMT, couplers and waveguide twist are integrated into a single block. The LO signal is injected from the bottom of the block through standard UG-387 waveguide flanges. Two mixer blocks, each equipped with a readout box, are mounted to the sides of the frontend assembly through custom design waveguide flanges. The IF readout circuits are integrated in the readout boxes. Because of the wideband specifications of the wSMA receivers, we have designed all these waveguide components to operate over a 40% fractional bandwidth.

II. Orthomode Transducer

An OMT is a passive device that separates an input signal into two orthogonal linearly polarized components. The double-ridge Bøifot type OMT design [2], [3] has been well studied for operating at millimeter and submillimeter wavebands. The wSMA OMT follows a similar design to that shown in [3]. The OMT junction, geometry of the side waveguides, the combiner, and the transformer design for the vertical waveguide were further optimized.

The HFSS model and photos of our WR4.3 OMT prototypes are shown in figure 2. The design is more compact than the previous designs in [2] and [3]. Our prototypes consist of four machined parts. To avoid the need to fabricate a deep hole for the vertical polarization port waveguide (P2), we replaced a long section of P2 by a straight split-block waveguide.

Fig. 2 CAD model and photos of our WR4.3 OMT prototype

We measured our OMT prototypes using a back-to-back configuration. We mated the input square waveguide ports of two OMTs and measured the S-parameters through the ports P1 and P2. This configuration gave us the worst-case scenario performance of the prototypes. Referring to figure 3, excellent performance over a full waveguide bandwidth was obtained. The total insertion loss of the two back-to-back OMTs is better than 1 dB across most of the band. Given the insertion waveguide length of each OMT is 30 mm, the intrinsic insertion loss of the OMT is very low. The measured polarization isolation is around 50 dB and the return losses of the rectangular waveguide ports are around -20 dB for both polarizations.

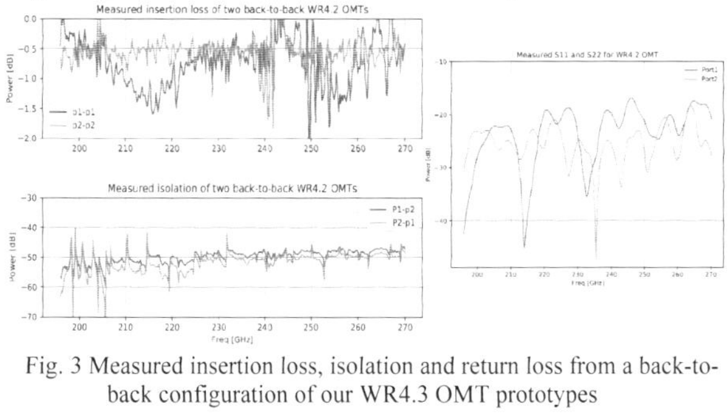

Fig. 3 Measured insertion loss, isolation and return loss from a back-to-back configuration of our WR4.3 OMT prototypes

---

[1] Center for Astrophysics | Harvard & Smithsonian
Email: lingzhen@cfa.harvard.edu

III. LO Waveguide Coupler

Chip-based directional couplers using three sets of probes connected by suspended striplines (bow-tie antennas) were proposed in [4], [5]. The original designs used three bow-tie antennas fabricated on individual quartz substrates. However, given the small chip sizes needed for operation at these wavelengths, it is difficult to install and align each individual in place, and this led to significant differences between the measurement and simulation data [5].

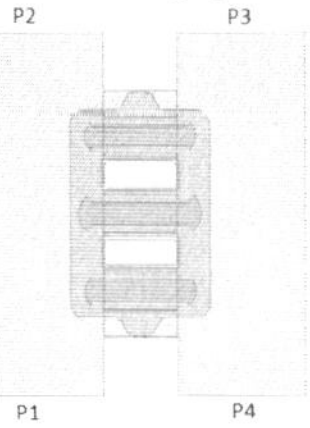

Fig. 4 schematic of waveguide chip coupler

Figure 4 shows the wSMA chip coupler design. Three bow-tie antennas were integrated on a single silicon chip, with two open slots in the center. This design allows easy and reliable chip installation. Our coupler prototype for WR4.0 band is shown in figure 5. Our measurement results agree very well with the simulations. As shown in figure 6, we achieved a flat coupling curved across a wide operation bandwidth. A paper showing the details of this coupler design is submitted to IEEE Transactions on Terahertz Science and Technology [6].

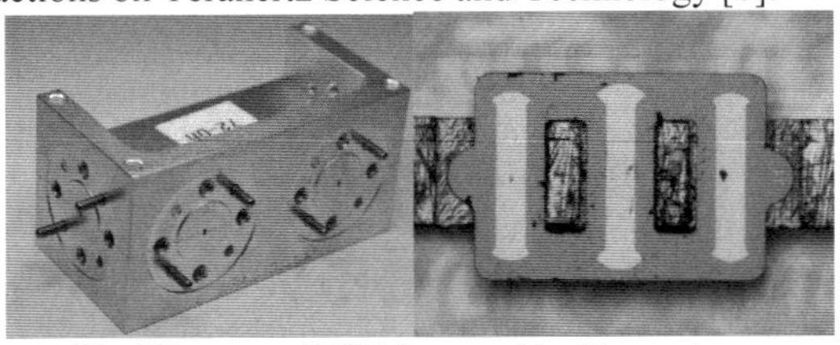

Fig. 5 Prototype of WR4.0 waveguide chip coupler

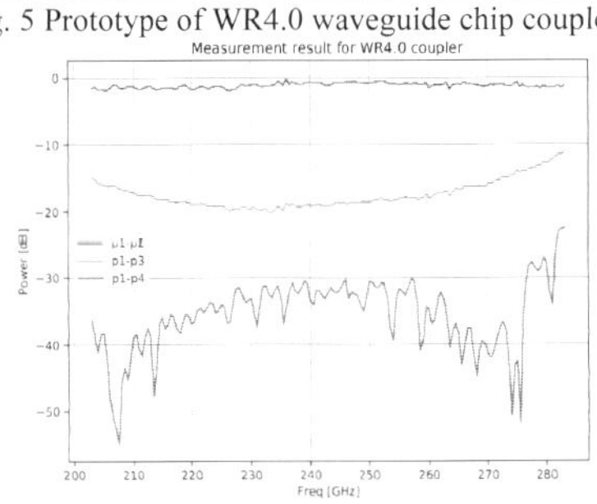

Fig. 6 measurement results of WR4.0 chip coupler

IV. Waveguide Twist

The function of a waveguide twist is to rotate the polarization plane of the electromagnetic wave propagating inside a waveguide system by a specified angle. In the wSMA frontend design, the waveguide twist rotates the orientation of one of the LO ports, allowing proper waveguide bending from the bottom of the receiver cartridge.

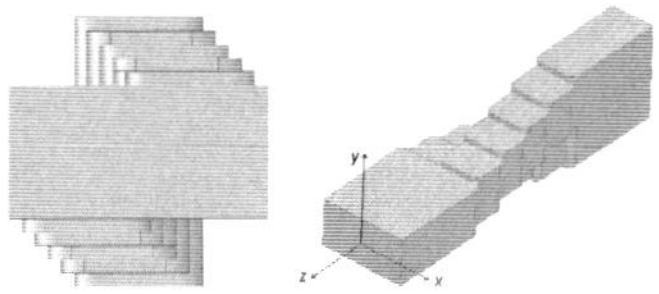

Fig. 7 CAD model of wSMA waveguide twist design

Our waveguide twist is a novel rectilinear design [7], which covers a full waveguide bandwidth. As shown in figure 7, the design achieves compactness and ease of fabrication by utilizing a series of quarter-wave transformer steps machined in split blocks. Figure 8 shows the prototypes of our WR3.4 waveguide twist. The top and bottom blocks are designed to be identical to reduce the fabrication costs. The measured S-parameters are shown in figure 9. The return losses are better than -25 dB and the insertion loss is about 0.6 dB across the entire waveguide band. This data includes a 12-mm long waveguide section from either side of the waveguide ports, leading to the actual twist itself. As a result, we believe that the actual performance of the twist should be even better.

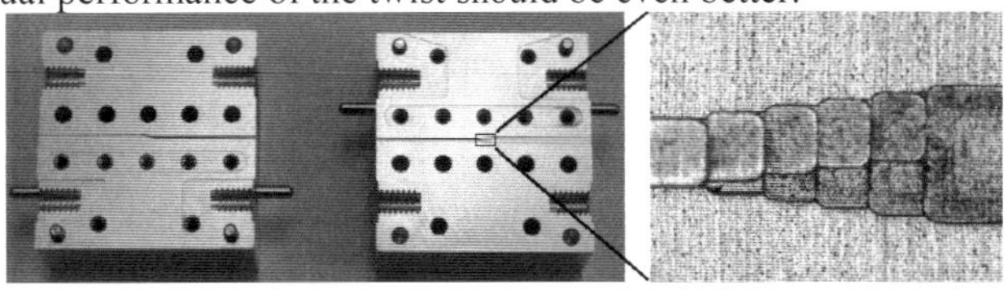

Fig. 8 Prototypes of wSMA waveguide twist

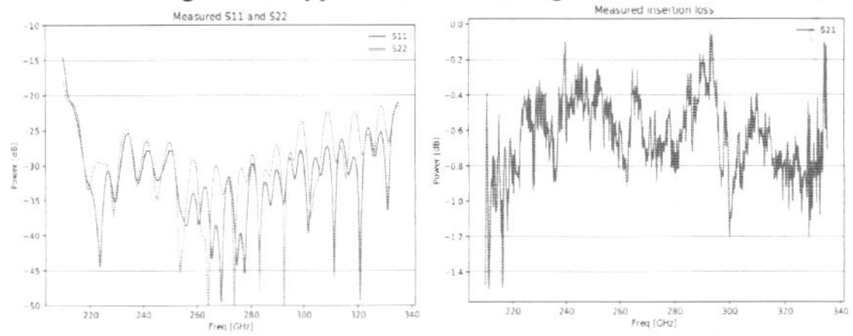

Fig. 9 Measurement results of wSMA waveguide twist

V. Frontend Integration

Using the components described in the above sections, we have designed an integrated frontend assembly, and Figure 10 gives the CAD model. It consists of an OMT, two couplers and one waveguide twist. The RF signal enters from the square input waveguide port. The LO signals are injected from the bottom of the assembly through standard UG-387 waveguide flanges.

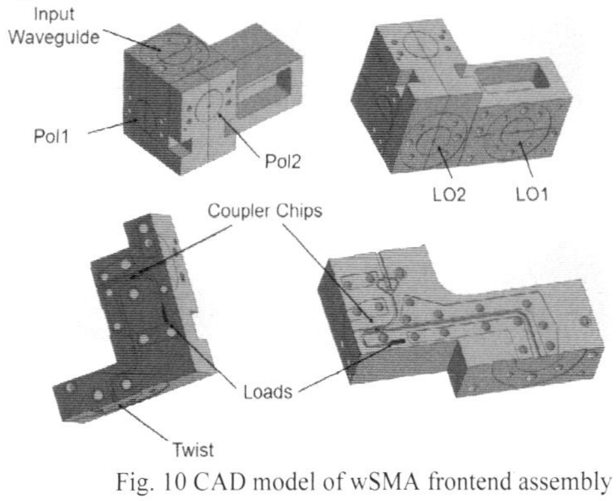

Fig. 10 CAD model of wSMA frontend assembly

As shown in figure 11, the frontend assembly will be integrated with a feedhorn and mixer blocks. The feedhorn is mounted on the top and the two mixer blocks and their associated readout boxes are mounted to the sides of the assembly through custom-designed waveguide flanges.

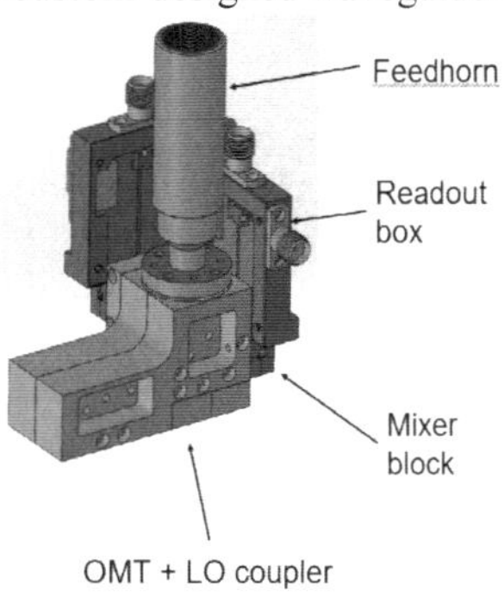

Fig. 11 CAD model of wSMA frontend assembly with feedhorn, mixer blocks and readout box

VI. CONCLUSION

A number of waveguide components have been developed for the wSMA frontend. The measurement results of the prototypes agree well with simulations and meet the requirements of the project. Further work frontend integration and system performance tests will follow.

REFERENCES

[1] Paul K. Grimes & Raymond Blundell & Patrick Leiker & Scott N. Paine & Edward C.-Y. Tong & Robert W.Wilson & Lingzhen Zeng. "Receivers for the wideband submillimeter array", Proceedings of the 31st International Symposium on Space Terahertz Technology (2020).

[2] Wollack, Edward & Grammer, W. & Kingsley, Jeffrey. THE BØIFOT ORTHOMODE JUNCTION. ALMA memo #425, (2002). http://library.nrao.edu/public/memos/alma/main/memo425.pdf

[3] Kamikura, Mamoru & Naruse, Masato & Asayama, Shin'ichiro & Satou, Naohisa & Wenlei, Shan & Sekimoto, Yutaro. "Development of a 385-500 GHz Orthomode Transducer (OMT)", Proceedings of the 19th International Symposium on Space Terahertz Technology (2008).

[4] P.K. Grimes at al. "GUBBINS: A Novel Millimeter-Wave Heterodyne Interferometer", Proceedings of the 20th International Symposium on Space Terahertz Technology (2009).

[5] Leech, J. & Yassin, Ghassan & Tan, Burak & Zhou, Y. & Garrett, John & Grimes, P.. "An SIS mixer based focal-plane array at 230 GHz", Proceedings of the 26th International Symposium on Space Terahertz Technology (2015).

[6] Lingzhen Zeng & Wei-Chun Lu & Paul K. Grimes & Tse-Jun Chen & Yen-pin Chang & C. Edward Tong & Ming-Jye Wang. " A Silicon-chip-based Waveguide Directional Coupler for Terahertz Applications", Submitted to IEEE Transactions on Terahertz Science and Technology (2020)

[7] L. Zeng & C. E. Tong & S. N. Paine & P. K. Grimes, "A Compact Machinable 90° Waveguide Twist for Broadband Applications," in IEEE Transactions on Microwave Theory and Techniques. (2020)

Measurements and Simulations of a 20 GHz Metamaterial Lens

Cassandra Whitton[1], Christopher Groppi[2], Philip Mauskopf[3], Jose Siles[4], and Adrian Tang[5]

Abstract— **In this paper, we present measurements of a prototype metamaterial flat lens. Flat lenses with short focal lengths are interesting due to their potential use in remote sensing for both space-based cubesat applications and larger form factors. Our metamaterial flat lens was manufactured by using 11 layers of RO3003 circuit board laminate with etched sub-wavelength-sized copper patterning. The copper patterning is designed in such a way as to maximize the transmittance of the lens while applying the correct phase shift across the lens plane to give the lens gaussian focal properties. The lens was measured by scanning a receiver horn through one axis of the image plane of a transmitting horn. This measurement demonstrated that the waist of the focused gaussian beam is 30% wider than ideal. It is suspected that this non-ideality is caused by phase error in the design process. We have created an electromagnetic simulation based on Fourier optics which will soon be able to characterize such non-idealities. Further measurements will be useful to fully characterize the lens's focal properties and determine how much loss it incurs.**

I. Introduction

CubeSats are an attractive prospective for those wishing to perform terahertz observations in space, due to the high atmospheric attenuation at these frequencies which makes ground-based observing difficult or impossible [7], and due to the prohibitively high costs of larger satellites. However, CubeSat missions come with their own set of design challenges, which particularly includes the requirements for low weight and small form factor [3]. As a partial solution to this, using metamaterial lenses as primary observing apertures in such systems can help ameliorate some of these challenges.

Metamaterials, which often involve the structured embedding of metal elements within dielectric substrates, and metamaterial lenses in particular, have recently seen much advancement and development into the millimeter wavelength regime [4], [5]. The lenses which have been created so far are both thin and lightweight compared to a conventional lens of equivalent f-number, freeing up weight budget and making it easier to place and stow the lens, if a deployable design is necessary or desirable. Furthermore, the design techniques introduced by Ref. [5] ensure that no anti-reflection coating is necessary to minimize reflection losses. Such lenses have been found to theoretically have less than half a dB of loss, which is significantly better than that of a Fresnel zone plate lens, which, while flat and light, can exhibit on the order of 3 to 4dB or more of loss [6].

Here we have created and tested a metamaterial flat lens which operates at 20 GHz. The lens we present here is intended to act as a low-frequency prototype to test our design procedure. We are also developing a complementary electromagnetic simulation to refine our design process and to more accurately determine the optical properties of our lenses before fabrication. A successful design procedure should allow us to experiment with more expensive high-frequency designs, operating at 600 GHz or even above 1 THz.

II. Lens Design

The lens is designed to operate at 20GHz by transforming a plane wave (with a flat phase-front) to the phase of a converging gaussian beam. This phase transformation is approximated by Refs. [5] and [2] as

$$\phi(r) = -\frac{\pi r^2}{\lambda R} \tag{1}$$

where λ is the operational wavelength, r is the distance on the lens plane from the lens center, R is the radius of curvature of the phase front, given by

$$R(f) = f + \frac{\left(\frac{\pi w_0^2}{\lambda}\right)^2}{f} \tag{2}$$

and w_0 is the waist of the focused beam at the focal plane. The focal length of the lens is given by f.

In the case of our lens, the diameter of the active area is 254 mm, and the focal length is 105 mm, making our lens an $f/0.41$ lens. These design parameters result in the phase transformation shown in the top plot of Fig. 1.

Given this phase transformation, we grid the surface of the lens vertically and horizontally into pixels, each of square dimensions $\lambda/10$, or 1.50 mm. Each of these pixels is assigned a single phase transformation value based on the above equations. In our case, each pixel has 10 metal layers, with 10 copper squares of metal, stacked on top of each other and separated by 11 surrounding dielectric layers. The dimensions of each square may be picked freely. Then, using techniques as described in Ref. [5], each pixel is optimized to give the desired phase transformation and maximum transmittance. These optimizations work by automatically

Submitted for review 28 February 2021.

This work was supported in part by the JPL SURP program. RO3003 material used in this work was provided gratis by Rogers Corporation.

[1]Cassandra Whitton is with the School of Earth and Space Exploration, Arizona State University, Tempe, AZ 85287, USA (e-mail: cassandra.whitton@gmail.com)

[2]Christopher Groppi is with the School of Earth and Space Exploration, Arizona State University, Tempe, AZ 85287, USA (e-mail: cgroppi@asu.edu)

[3]Philip Mauskopf is with the School of Earth and Space Exploration, Arizona State University, Tempe, AZ 85287, USA (e-mail: philip.mauskopf@asu.edu)

[4]Jose Siles is with the Jet Propulsion Laboratory, Pasadena, CA 91109, USA (e-mail: jose.v.siles@jpl.nasa.gov)

[5]Adrian Tang is with the Jet Propulsion Laboratory, Pasadena, CA 91109, USA (e-mail: adrian.j.tang@jpl.nasa.gov)

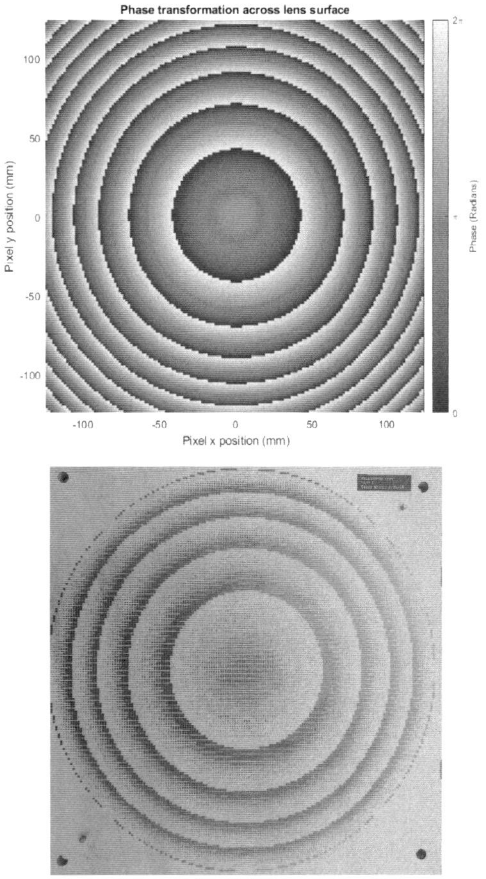

Fig. 1. The designed lens phase transformation (above) is compared with one layer of the manufactured lens (below). The layer's pattern is made up of thousands of copper squares.

tweaking the dimensions of each metal square until the desired conditions for that pixel are met. Because each pixel is treated independently from each other pixel, this is relatively computationally simple, as it only requires the optimization of around 5 free parameters (one for each layer of the lens, divided by 2 due to symmetry across its center) per pixel.

The lens was manufactured on RO3003 circuit board laminate with 760 mm thickness, with 1 ounce copper cladding. Due to manufacturing tolerances, the metal squares were constrained to be no smaller than 200 um in dimension, with at least 200 um between adjacent squares. A single layer of the manufactured lens is shown in the bottom half of Fig. 1. The layers were then stacked together, as shown in Fig. 2, using alignment holes that were drilled into the laminate layers during manufacturing. The completed lens is roughly 0.59 cm thick.

Though not directly relevant to this experiment, the lens is designed in such a way that it may be scaled from 20 GHz to 600 GHz. In doing so, the layer thicknesses and metal square sizes would be reduced by a factor of 30. Though this places much stricter tolerances on the manufacturing process,

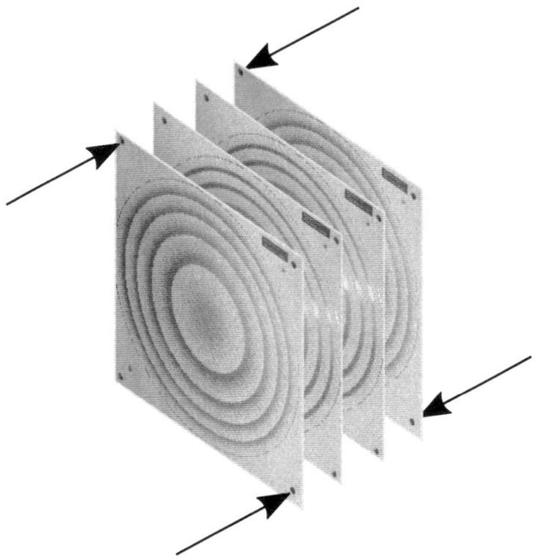

Fig. 2. Multiple layers of the lens are stacked together, aligned with guide-pins, and held together through boltholes at the corners. While this figure demonstrates only 4 layers, our lens has 11.

we have confirmed that the ASU Flexible Display Center is capable of manufacturing

III. Experimental Setup

In order to test that the lens operates correctly, we performed a simple image-plane measurement. The physical setup for the experiment is shown in Fig. 3. To accomplish this, the transmitter (Tx) and receiver (Rx) were each placed 2 focal lengths away from the lens. We do this because placing the Tx $2f$ away from the lens causes its image to be $2f$ from the lens on the other side, as shown by the well-known lens equation:

$$\frac{1}{f} = \frac{1}{d_o} + \frac{1}{d_i} \tag{3}$$

Here, d_o is the distance of the object from the lens and d_i is the distance of the image from the lens. Thus, the receiver directly measures the image of the transmitter. In addition, the resultant absolute magnification of the image is 1.

The transmitter and receiver each consist of a K-band pyramidal horn antenna with nominal 14 dBi gain coupled to a WR-42 waveguide. The transmitter is fed by a signal generator emitting a 20 GHz tone at -20 dBm. The receiver was connected to a power meter. The power meter only nominally operates up to 18 GHz; however, we tested that the power meter responded linearly to power input at 20 GHz. Therefore, while the absolute measurements of the power meter were likely incorrect, we are confident in the relative power measurements that it provided.

Once the transmitter's position was set, the receiver was scanned manually through the image plane until the point of maximum power reception was found. This was used as the zero-point for the measurement. The receiver was then manually moved up and down through the image plane in increments of a couple of millimeters. At each stopping point, the height of the receiver relative to its zero-point was

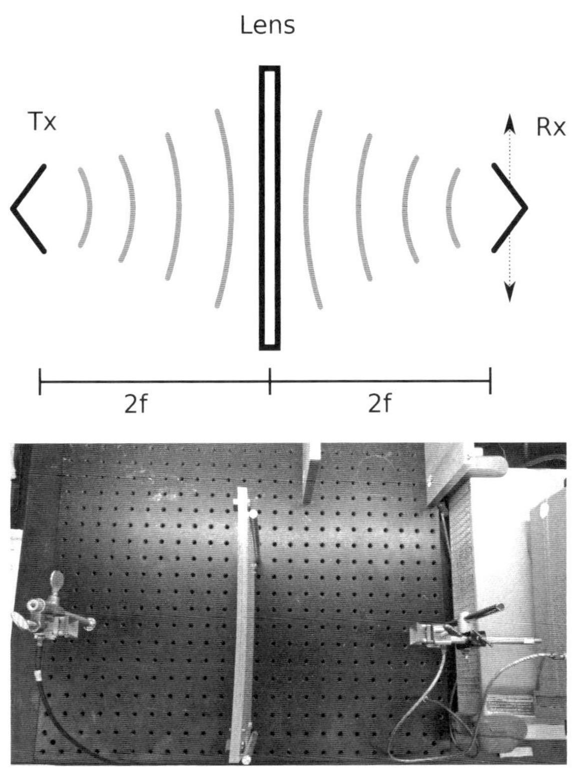

Fig. 3. A diagram of the test setup is display, with the realized setup displayed below it. The receiver and transmitter are both placed two focal lengths from the lens, such that the receiver sees an un-magnified image of the transmitter.

recorded, along with the power measurement from the power meter.

IV. Electromagnetic Simulation

In order to further verify the accuracy of our lens design process, we have implemented an electromagnetic simulation which can propagate simulated beams. The simulation uses Fourier optics, implemented as described in Ref. [1].

We have demonstrated the ability of our simulation to handle lens beam transformations. In Fig. 4, we have plotted a simulation of the test setup described in the previous section. As we expected, we can see the beam diverge from the transmitter horn before being refocused to the receiver by the lens in the center.

We have not yet simulated the beam waist that we measured in our experiment and plotted in Fig. 5. However, in the future, this will serve as an important check that our design process works the way we think it does. It will also allow us to test the optical properties of our designs before we manufacture them.

V. Results

Plotted with red circles in Fig. 5 is the result of the image plane measurement as described above. The magenta line is the best gaussian beam fit to the measurement.

Plotted in blue is what we would expect to measure if the lens were acting as an ideal lens. This was calculated

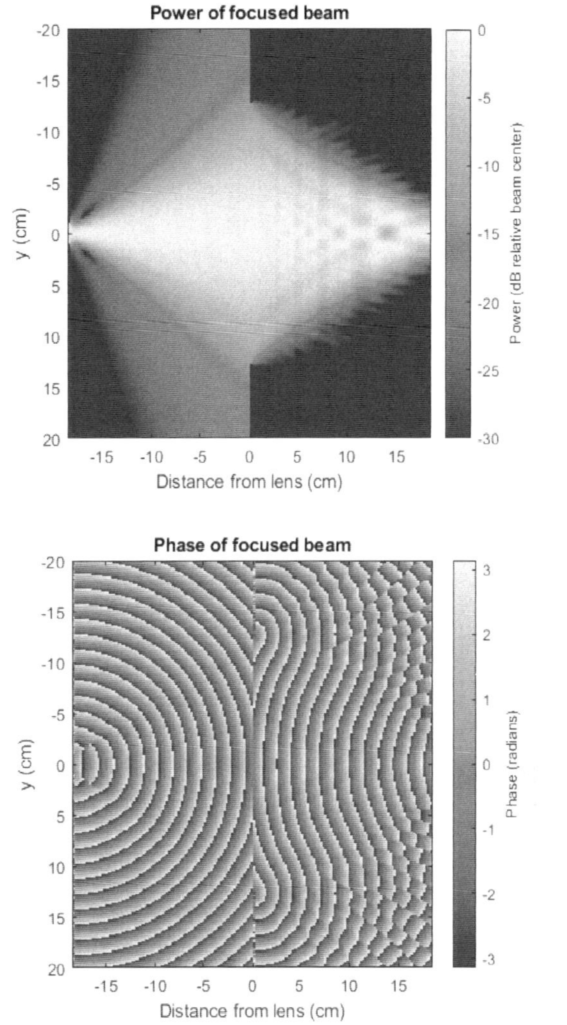

Fig. 4. Here is an electromagnetic simulation of our experimental test setup, plotted in magnitude and phase. On the left, the transmitter is emitting a roughly gaussian beam. On the right, the lens focuses the beam inward toward the receiver.

using the properties of the horn antennas, which we believe have beam waists of 8.4 mm at 20 GHz. By convolving the gaussian beam shapes of the two horns together, as described in Ref. [2], we expect the measured image to have an effective gaussian waist of approximately 11.9 mm.

The effective beam waist of the gaussian beam fit was 15.4 mm, which was about 30% wider than ideal. This indicates that the lens focuses properly, but does not focus as ideally as we expected it to. Future work will investigate the reasons for this non-ideal behaviour.

VI. Conclusion

We have successfully demonstrated the focusing abilities of our 20 GHz metamaterial flat lens. However, the gaussian beam focus is 30% wider than ideal.

Further testing would be necessary to determine whether this is due to manufacturer tolerance error, phase error in

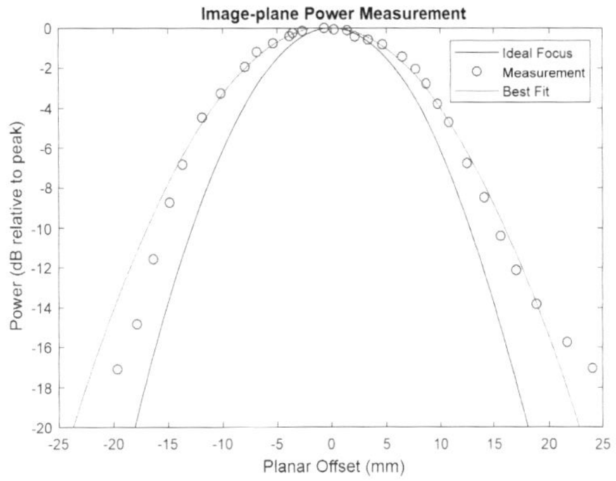

Fig. 5. Here is the plotted measurement of the image-plane scan compared to the expected beam measurement for an ideally focusing lens. The measured beam-width is about 30% wider than ideal.

the initial design, or an inaccurate estimate of the beam waists of the transmitter and receiver horns which were used. The simulation of the designed lens, which we are currently implementing, with its phase error included, could reveal whether or not the phase error was the cause. We also believe that a full near-field measurement of the lens would help to quantify the lens's performance in more meaningful ways.

In any case, we have demonstrated that our design process works well enough to accomplish an acceptable focus with our lens. We have also performed preliminary tests to demonstrate that our electromagnetic lens simulation is working as intended, which will help significantly in designing and testing future lenses. Using these results as a foundation, we will in future work scale the lens to higher frequencies. Doing so would allow their eventual application in CubeSat-based terahertz observations.

ACKNOWLEDGEMENTS

We would like to thank Rogers Corporation for providing us with RO3003 circuit board material free of charge, and the JPL SURP program for their financial support.

REFERENCES

[1] Kristina K Davis, Willem Jellema, Stephen JC Yates, Christopher E Groppi, Jochem JA Baselmans, and Andrey Baryshev. Analysis techniques for complex field radiation pattern measurements. In *Millimeter, Submillimeter, and Far-Infrared Detectors and Instrumentation for Astronomy IX*, volume 10708, page 107082R. International Society for Optics and Photonics, 2018.

[2] P.F. Goldsmith. *Quasioptical Systems: Gaussian Beam Quasioptical Propogation and Applications*. IEEE Press Series on RF and Microwave Technology. Wiley, 1998.

[3] Eva Peral, Eastwood Im, Lauren Wye, Simon Lee, Simone Tanelli, Yahya Rahmat-Samii, Stephen Horst, Jim Hoffman, Sang-Ho Yun, Travis Imken, et al. Radar technologies for earth remote sensing from cubesat platforms. *Proceedings of the IEEE*, 106(3):404–418, 2018.

[4] G Pisano, C Tucker, PA R Ade, P Moseley, and MW Ng. Metal mesh based metamaterials for millimetre wave and thz astronomy applications. In *Millimeter Waves and THz Technology Workshop (UCMMT), 2015 8th UK, Europe, China*, pages 1–4. IEEE, 2015.

[5] Giampaolo Pisano, Ming Wah Ng, Fahri Ozturk, Bruno Maffei, and Vic Haynes. Dielectrically embedded flat mesh lens for millimeter waves applications. *Applied optics*, 52(11):2218–2225, 2013.

[6] JM Rodriguez, Hristo D Hristov, and Walter Grote. Fresnel zone plate and ordinary lens antennas: Comparative study at microwave and terahertz frequencies. In *Microwave Conference (EuMC), 2011 41st European*, pages 894–897. IEEE, 2011.

[7] David M Slocum, Thomas M Goyette, and Robert H Giles. High-resolution terahertz atmospheric water vapor continuum measurements. In *Terahertz Physics, Devices, and Systems VIII: Advanced Applications in Industry and Defense*, volume 9102, page 91020E. International Society for Optics and Photonics, 2014.

Development of readout electronics for SIS photon counting detectors

Hiroshi Matsuo[1,2], Hajime Ezawa[1,2], Ryohei Noji[3], Saho Kawahara[3]

Photon counting detectors for terahertz frequencies will open a new frontier in terahertz astronomy by measuring photon statistics and applying to intensity interferometry. To count large number of terahertz photons, we work on SIS (or STJ) photon detectors. In this presentation we present fast readout cryogenic electronics with GHz bandwidth made of semiconductor circuits for the SIS photon detectors.

Single terahertz photon creates single quasi-particle current through SIS junction by photon assisted tunneling. We have achieved the SIS leakage current of the order of pA, whose photovoltaic signal can be readout by cryogenic FETs with low gate leakage and capacitance. We selected two types of Gallium Arsenide FETs with p-n junction gates (GaAs-JFET and Junction pHEMT) for this purpose.

We have evaluated both types of FETs with various gate sizes at cryogenic temperature at 4 K, most of which show good I-V characteristics without serious anomalies such as kink or hysteresis. Transconductances of these FETs are similar or improved at cryogenic temperatures compared to the ones at 300 K. The gate leakage of GaAs-JFET have been measured as a function of operational temperature, which shows the gate leakage are lower than 100 fA at temperature less than 100 K. The gate leakage of Junction pHEMT (JPHEMT) shows similar trend except that the gate leakage shows large excess current at drain voltages larger than 2 V, and we must limit drain voltage less than 1.5 V.

Figure 1 shows an example of I-V characteristics of two types of FETs measured at 4.2 K. Compared to GaAs-JFET, JPHEMT apparently shows better performance with larger transconductance and drain resistance which are appropriate for our purpose. Slight anomalies are observed at larger drain voltage for JPHEMT, which corresponds to the anomalous gate leakage described above.

We plan to use two-stage source followers, one at 0.8 K and another at 4 K, to decrease output impedance down to 50 Ohm to feed to SiGe low noise amplifiers. The power dissipation is limited to 100 μW at 0.8 K, and the first stage source follower decreased the output impedance to an order of kΩ followed by the second source follower to achieve lower output impedance.

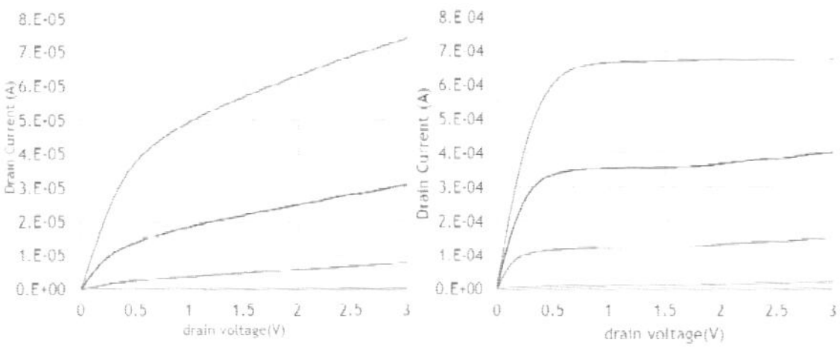

Fig. 1. I-V characteristics of GaAs-JFET and Junction pHEMT.

For detection of single terahertz photon, low gate capacitance less than 10 fF is required to obtain signal above the FETs' voltage noise. The evaluation of the gate capacitance is also underway.

The detector design themselves must be optimized for the low capacitance readout electronics. Since previous design uses a relatively large choke filter and bonding pads. Capacitance of all these structures should be optimized, which is under study and will be discussed in the presentation.

REFERENCES

[1] H. Matsuo, H. Ezawa, "Advantages of Photon Counting Detectors for Terahertz Astronomy", *J. Low Temp. Phys.* 184, 718-723 (2016).

[2] H. Ezawa, H. Matsuo, M. Ukibe, G. Fujii, S. Shiki, "Studies on Terahertz Photon Counting Detectors with Low-Leakage SIS Junctions", *J. Low Temp. Phys.* 194, 426-432 (2019).

[1] National Astronomical Observatory of Japan, Mitaka, Tokyo 181-8588, Japan.

[2] Department of Astronomical Science, The Graduate University for Advanced Studies, SOKENDAI, Tokyo 181-8588, Japan.

[3] Toho University, Funabashi, Chiba 274-8510, Japan.

Development of 109-pixel NbTiN-Al MKID array for the 100-GHz band astronomical observations

Y. Murayama[1,2], T. Nitta[1,3], M. Nagai[2], R. Hikawa[1], R. Suzuki[1], W. Shan[2], H. Matsuo[2], A. Miyachi[2], M. Kroug[2], S. Ezaki[2], Y. Sekimoto[4], T. Noguchi[5], M. Naruse[6], N. Kuno[1,3], and N. Nakai[7]

We present the development of a 109-pixel microwave kinetic inductance detectors (MKIDs) array for ground-based 100-GHz band astronomical observations. Each pixel of this array consists of an antireflection-coated Si lenslet and a double-slot antenna coupled coplanar waveguide (CPW) $\lambda/4$ -resonator MKID. To increase the optical efficiency, we have introduced the NbTiN-Al hybrid MKIDs [1] to our array. The NbTiN transmission line and ground plane are lossless for both the readout frequency of 5-GHz band and the 100-GHz band radiation.

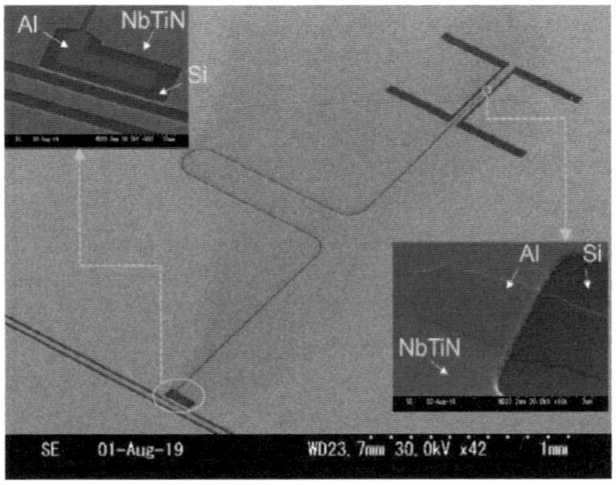

Fig. 1. SEM image of NbTiN-Al hybrid MKID. The center strip of resonator CPW (3 μm width) is Al and the ground plane and transmission line (56 μm width) are made of NbTiN.

The designed length and width of Al center strips are 4.17 - 4.53 mm and 3 μm. The CPW $\lambda/4$-resonator is weakly coupled to the transmission line via a wide coupler at open-end. At short-end, a double-slot antenna optimized for 100-GHz band is connected to the resonator (Fig. 1). We attached a short-circuit stub to antenna feed point for impedance matching [2], based on HFSS simulation.

The MKIDs are distributed over the entire 3-inch Si substrate. The Si substrate was cleaned by buffered HF to remove the oxide layer, just before each metal sputtering. At the center of the 3-inch substrate, the thickness of NbTiN and Al are 237 nm and 42 nm, respectively. After an exposure on 1 μm thick positive resist PFI-38A9 by a contact aligner and the development, the films were etched by ICP-RIE using CF_4 and O_2 gases for NbTiN and by wet etching for Al. To understand the NbTiN film we did the R-T measurement of a meander pattern NbTiN cooled to ~ 4.2 K and obtained T_c = 13 K, $2\Delta_0/h \cong 0.95$ THz, and $\rho_{20K} = 157$ μΩcm.

To evaluate the optical properties of the hybrid MKIDs, the array was installed in a Nb holder and cooled to 66 mK using a dilution refrigerator. The optical response and the optical efficiency were measured by 300 K and 77 K loads. We confirmed the optical response of 104 pixels. The measured optical efficiency is about 3 times higher than that of the Al single layer MKID array. Fourier transform spectrometer was used for checking the array frequency response. The measured center frequency and bandwidth are 95 GHz and 16 GHz, respectively. This result is in good agreement with the expected response which is determined by the Al superconducting gap and the 120-GHz low-pass filter used in the cold optics.

We also discuss some dark measurement results and the noise measurement under the optical loading.

References

[1] S. J. C. Yates, et al. "Photon noise limited radiation detection with lens-antenna coupled microwave kinetic inductance detectors." *Applied Physics Letters*, 99.7 (2011): 073505.

[2] L. Ferrari, et al. "Antenna coupled MKID performance verification at 850 GHz for large format astrophysics arrays." *IEEE Transactions on Terahertz Science and Technology*, 8.1 (2017): 127-139

[1] Division of Physics, University of Tsukuba, 1-1-1 Tennodai, Tsukuba, Ibaraki 305-8571, Japan.

[2] Advanced Technology Center (ATC), National Astronomical Observatory of Japan (NAOJ), 2-21-1 Osawa, Mitaka, Tokyo 181-8588, Japan.

[3] Tomonaga Center for the History of the Universe, University of Tsukuba, 1-1-1 Tennodai, Tsukuba, Ibaraki 305-8571, Japan.

[4] Department of Space Astronomy and Astrophysics, Institute of Space and Astronautical Science (ISAS), Japan Aerospace Exploration Agency (JAXA), 3-1-1 Yoshinodai, Chuo-ku, Sagamihara City, Kanagawa 252-5210, Japan.

[5] The University of Electro-Communications, 1-5-1 Chofugaoka, Chofu, Tokyo 182-8585, Japan.

[6] Graduate School of Science and Engineering, Saitama University, 255 Shimo-Okubo, Sakura-ku, Saitama, Saitama 338-8570, Japan.

[7] Department of Physics, School of Science and Technology, Kwansei Gakuin University, 2-1 Gakuen, Sanda, Hyogo 669-1337, Japan.

High Power Amplifier Modules from 110 to 200GHz

Theodore Reck[1], Zach Griffith[2], and Eric Bryerton[1]

Terahertz multiplier-based local-oscillator chains offer the highest spectral purity but, because of low efficiencies of the multipliers, require large DC powers to support the millimeter-wave amplifiers that drive them. In addition, the engineering challenge of aligning multiple stages of narrow-band varactor multipliers makes high-power broad-bandwidth sources difficult. Moving the power-amplifier to a higher frequency in the chain can alleviate these issues.

Utilizing Teledyne's InP HBT MMICs process we have developed a high-power amplifier module operating from 115-130GHz [1]. This device produces over 180mW with only 1mW input power and consumes 3.5W of DC power. In conjunction with low-power, broad-band multipliers, this amplifier can replace the first stages of high-power varactor multipliers. This amplifier-based system would consume a quarter of the DC power while still providing 19% bandwidth. In addition to this 115-140 GHz device, other amplifier modules are being developed at 155-180 GHz and 180-210GHz which will be presented at the conference.

For higher-power applications, power-combined modules are also under development. These use a hybrid-coupler corporate combiner architecture to combine four MMICs into a single package. Measurements of a 4-way power combined 130GHz amplifier will be shown at the conference.

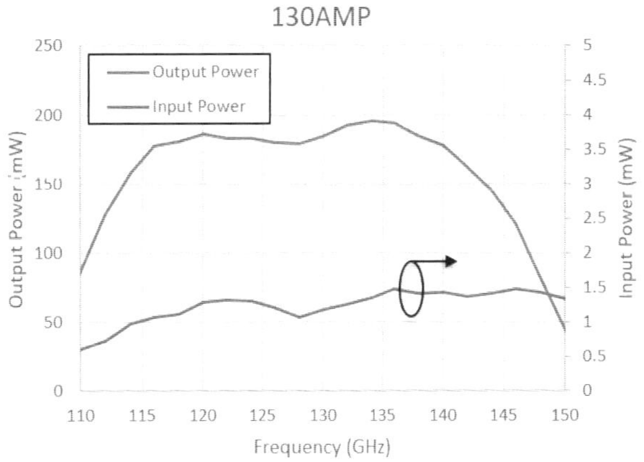

Fig. 1. Measured data from the packaged 115-140 GHz InP HBT MMIC. DC power consumption is 3.5W.

References

[1] Z. Griffith, M. Urteaga and P. Rowell, "A 140-GHz 0.25-W PA and a 55-135 GHz 115-135 mW PA, High-Gain, Broadband Power Amplifier MMICs in 250-nm InP HBT," *2019 IEEE MTT-S International Microwave Symposium (IMS)*, Boston, MA, USA, 2019, pp. 1245-1248.

[1] Virginia Diodes Inc., Charlottesville, VA 22902 USA.
[2] Teledyne Scientific Company, Thousand Oaks, CA, USA.

Reliability study of THz Schottky mixers and HBV frequency multipliers for space applications

Vladimir Drakinskiy[1], Josip Vukusic[1], Daniel Heinerås[2], Peter Sobis[3], Vaclav Valenta[4], Marie-Genevieve Perichaud[4], Fernando Martinez Martin[4] and Jan Stake[1]

Abstract—We report status and current results of the preliminary reliability study on 300 GHz InP heterostructure barrier varactor diode multipliers and 1200 GHz GaAs Schottky diode mixers. Both types of diodes are monolithically integrated with circuits and were processed on 3" InP and GaAs wafers respectively using established III-V processing. We will present results on thermal step-stress tests up to 300°C, indicating the operational temperature limitations of the devices. Also, the analysis of the accelerated lifetime testing (1000h) will be discussed.

INTRODUCTION

There is a need for efficient and reliable THz sources and heterodyne receivers operating in the sub-millimetre wave band above 300 GHz for future space science missions and earth observation instruments. The sub-millimetre wave regime allows the study of different meteorological phenomena such as water vapor, ice and water content in clouds, and ice particle sizes and distribution, which are important parameters for the hydrological cycle of the climate system and the energy budget of the atmosphere or even explore outer space. Several space bound exploratory missions have carried instrumentation operating in this frequency range. Besides of high level of integration required for operation at THz range, the components utilized for space instruments, must be highly reliable, i.e. perform as expected throughout its planned lifetime. It is therefore important to conduct reliability tests on device/component level to ensure projected performance in future missions. This work describes the preliminary reliability testing of 300 GHz heterostructure barrier varactor diode frequency multipliers [1] and 1200 GHz Schottky diode mixers [2] fabricated on 3" wafers.

The reliability tests start out with step-stress tests of temperature, dc bias and RF. These tests will give us the limits T_{max}, V_{max} and RF_{in} which will be used as boundary conditions for the followed accelerated lifetime tests. Fig 1 shows the progression of the current-voltage relationship of a heterostructure barrier varactor during a temperature step-stress test ranging from 120 °C to 300 °C.

The overall aim of these tests is to fully understand the limitations of our devices so that we can improve, through augmented fabrication, as well as better predict their behavior to prove their reliability as state-of-the-art, THz workhorse devices.

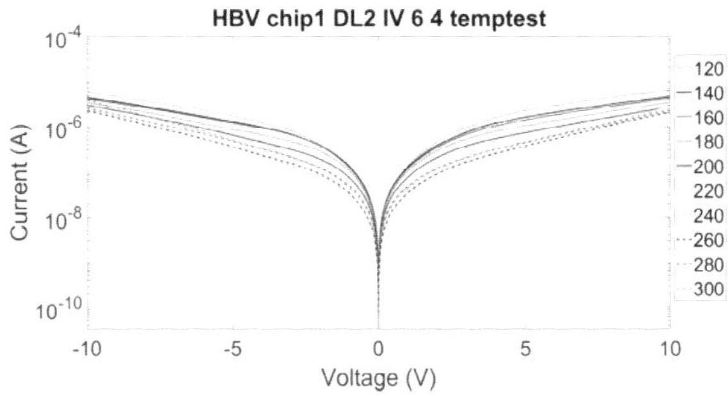

Fig. 1. 24h temperature step-stress test result (120-300 °C) on heterostructure barrier varactor diodes chips fabricated at MC2, Chalmers.

ACKNOWLEDGEMENT

This work was performed in part at Myfab Chalmers in a project funded by European Space Agency within General Support Technology Program – "HBV and Schottky components for Space".

REFERENCES

[1] J. Stake, A. Malko, T. Bryllert, and J. Vukusic, "Status and prospects of high-power heterostructure barrier varactor frequency multipliers," Proc. IEEE, vol. 105, no. 6, pp. 1008–1019, Jan. 2017.

[2] P. Sobis, V. Drakinskiy, N. Wadefalk, P. Å. Nilsson, A. Hammar, D. Nyberg, A. Emrich, H. Zhao, T. Bryllert, A. Y. Tang, J. Schleeh, J. Grahn, and J. Stake, "SWI 1200/600 GHz Highly Integrated Receiver Front-Ends", Proceedings of the 36th ESA Antenna Workshop on Antennas and RF Systems for Space Science, ESA/ESTEC, Noordwijk, The Netherlands, October 2015

[1] Terahertz and Millimetre wave laboratory, MC2, Chalmers University of Technology, Gothenburg, SE-412 96, Sweden.
[2] Wasa Millimeter Wave AB, Gothenburg, SE-412 96, Sweden.
[3] Omnisys Instruments AB, Västra Frölunda, SE-421 32, Sweden.
[4] European Space Agency – ESA/ESTEC, Noordwijk, PO Box 299 NL-2200 AG, The Netherlands

Performance Comparison of Fabricated 90 Degree Waveguide Twists using Direct Milling

Usman Shehryar[1]

Waveguide (WG) twists are required for the interconnections of waveguide systems, polarization splitter, OMT etc. These devices served the purpose to rotate the input signal to some angle to interconnect the output device. Commonly, this axial rotation is 90 degrees. Commercially, waveguide twists exist with smooth transition in the twist but they are larger in size with length equal to some multiple of the waveguide lengths.

At the THz and millimeter wave frequencies, the instruments have the requirement to be of compact size. This makes the fabrication difficult due to the size. Electroforming fabrication is expensive and electro-erosion technique is not accurate due to roughness. So a cost effective and accurate fabrication technique is to use the direct milling tools to implement a compact size WG twists. Previously, various single step WG twists of various geometries are demonstrated with good practical results. Their transforming unit is a single step WG of various geometries contains sharp corner cuts and ridges. Although they are compact size and broadband but they are much sensitive to the geometrical dimensions and have tight fabrication tolerances.

This paper has investigated two designs, one is 45 degree rotated double rigid WG twist [1] and second is WG twist with square and step ridges [2]. They are fabricated with milling tools of different radius and their performances are compared with their theoretical ideal geometries. WR-05 waveguides are used as input and output waveguides with corresponding single step WG twists are designed and simulated in the HFSS software to see the tolerances for the fabrication with different milling tools. With direct milling, some corners inside the twist will become "chamfered" with radius of the milling tool. The performance penalties due to deviation of the twist geometry from the ideal layout geometry shows that for 45-degree twist, a milling tool not more than radius 0.04mm, will keep the performance of the ideal geometry. For radius 0.05mm and greater, the performance starts to change at the lower frequency and has degradation of about 2.5 dB in the return loss at these frequencies for radius 0.15mm. (1.94% decrease in fractional bandwidth). Similar case is with the other W.G twist geometry. So, these twists are compact, broadband and easily fabricated with milling tools with appropriate radius size without effecting the performance compared to the ideal geometry.

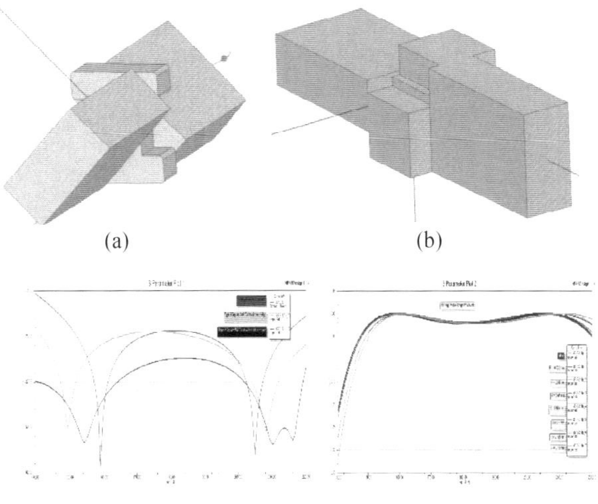

(a) (b) (c) (d)

Fig. 1. (a) 45 degree rotated double rigid WG twist (b) WG twist with square and step ridges degree rotated double rigid WG twist (c) S11-Parameters for the above two designs (d) S21-parametrs for the above two designs.

REFERENCES

[1] V.R. Litvinov, L. A. Rud, and E. A. Sverdlenko, "Compact 90° Twists Into Rectangular Waveguides", ISSN 0735 2727, Radioelectronics and Communications Systems, 2010, Vol. 53, No. 3, pp. 162–166. © Allerton Press, Inc., 2010. DOI: 10.3103/S0735272710030064.

[2] Miao Sun, Yong Xu, Tinghui Peng, Yafen Shang, Qiang Zhen, " Design of a Compact and Broadband 90-degree Waveguide Twist", 2018 IEEE International Vacuum Electronics Conference (IVEC), 24-26 April 2018, IEEE Xplore. **DOI**: 10.1109/IVEC.2018.8391621.

[1]MC2 Department, Chalmers University of Technology, SE412 96, Gothenburg, Sweden.

Tuning of superconducting Ti and Ti/Au bilayer films for transition-edge sensors

W. Zhang[1,2], Z. Wang[1,2], P.Z. Li[1,2,3], Y. Gen[1,2,3], J.Q. Zhong[1,2], W. Miao[1,2], Y. Ren[1,2], K.M. Zhou[1,2], Q.J. Yao[1,2] and S.C. Shi[1,2]

Superconducting transition-edge sensors (TES) can detect the electromagnetic radiations in a wide range from millimeter, optical to γ-ray. The critical temperature (T_c) of superconducting films is very important since it determines the key parameters of TESs, including noise equivalent power (NEP), energy resolution, response time. Ti film is a good candidate, and its T_c can be tuned by changing the thickness, bilayer or baking.

In fact, Ti thin films have been widely used in many applications, such as terahertz bolometers and optical photon detectors[1]. We have developed a 8×8 TES array with a NEP of 5×10^{-17} W/√Hz, and optical single photon detectors based on Ti TESs with a detection efficiency of 40%. Here we further tune the T_c of Ti thin films with magnetic field and baking in air atmosphere.

We fabricated pure Ti or Ti/Au films with different thicknesses on silicon substrate by electron beam evaporation [2]. Resistance measurement was carried out in a commercial physical property measurement system (PPMS) equipped with an adiabatic demagnetization refrigerator (ADR), and the base temperature is about 100 mK. We obtained a T_c of 331 mK for a 36.8 nm thick Ti film and it is reduced to 215 mK for a 29.5 nm one. That means it's possible to tune the T_c of Ti film by changing its thickness. However it's hard to decrease the T_c further since the superconductivity disappears while reducing its thickness below 25 nm. The transmission electron microscopy (TEM) showed that titanium films grown in our condition were multi-crystalline, confirmed by x-ray diffraction spectrum. We also measured the chemical composition with a ESD and found that there was 2~3% oxygen atoms throughout the whole thickness, and the oxygen composition was increased to 9% for such titanium films without superconductivity.

We then studied the effect of magnetic field on T_c of Ti films (see Fig.1). With the increase of magnetic field from zero to 600 Gauss, T_c was decreased from 331mK to 210 mK for 36.8-nm Ti film. The transition width nearly keeps constant, that means the temperature sensitivity (α=T/R×dR/dT) has no change within the range of measured magnetic field.

[1]Purple Mountain Observatory, CAS, China.
[2]Key Lab of Radio Astronomy, CAS, China.
[3]University of Science and Technology of China, China.

NOTES:

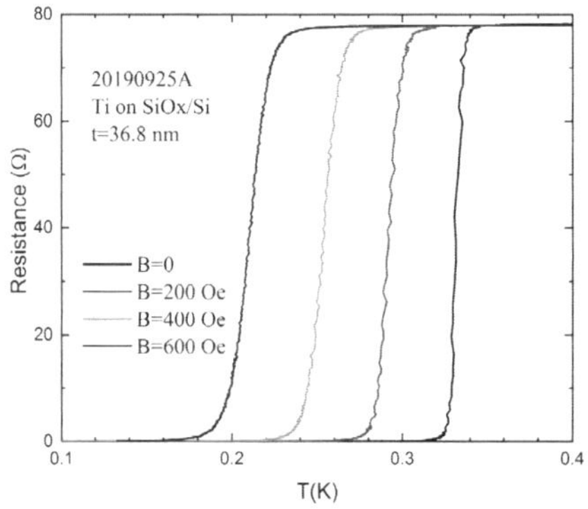

Fig. 1. Resistive transition of 36.8nm thick titanium film with different magnetic fields.

Finally, we tried to tune T_c by baking Ti films at 100°C for time between 5 and 30 hours in air atmosphere. The change in T_c (ΔT_c) was measured by finding a T_c for each sample, baking the sample and measuring its T_c again. For a 31.4 nm Ti film fabricated on 1550nm dielectric mirror, T_c was decreased from 304 mK to 235 mK when baking it for 10 hours (ΔT_c of 23%). In comparison, for 48/5nm Ti/Au bilayer T_c was decreased from 340 mK to 291 mK when baking it for 30 hours, about 14% change in T_c. The results indicate that it's possible to tune the T_c of both pure Ti and Ti/Au bilayer films, although 5nm Au film is a good protection layer for its underneath Ti film. We are going to tune the T_c of Ti TES and will present it in detail.

REFERENCES

[1] D. Fukuda, G. Fujii, T. Numata, et al, "Titanium-based transition-edge photon number resolving detector with 98% detection efficiency with index-matched small gap fiber coupling, Opt. Express 19(2),870, 2011.

[2] W. Wang, W. Zhang, W. Miao, D. Liu, J.Q. Zhong, and S.C. Shi, "Electron-beam evaportated superconducting titanium thin films for antenna-coupled transition edge sensors," IEEE Trans. Appl. Supercnd. 28(4), 2100204 (2017), DOI: 10.1109/TASC.2018.2799418.

A 200 GHz cloud radar multiplexing antenna

Stuart Froud, Manju Henry, Peter G. Huggard, Duncan A. Robertson, Soe Min Tun, Hui Wang & Richard Wylde

A team consisting of the STFC Rutherford Appleton Laboratory (RAL), Thomas Keating and St Andrews and Leicester universities are building and demonstrating a 200 GHz, 1.5 mm wavelength, cloud profiling radar, establishing the basis for the future operation of a similar Earth observing radar in space. The pulsed Doppler radar is called GRaCE: G-band Radar for Cloud Evaluation, and is grant funded by the UK Space Agency through the UK Centre for Earth Observation Instrumentation [1].

The small wavelength of a 200 GHz space radar will provide enhanced global information on the distribution of small droplets in the atmosphere. When operated in tandem with existing lower frequency space radars, the dual wavelength observations will enable atmospheric scientists to better characterize the microphysical properties of hydrometeors in water and ice clouds. Such information is needed to improve the accuracy of societally important numerical weather prediction models

A model of the GRaCE hardware is set out below in Figure 1.

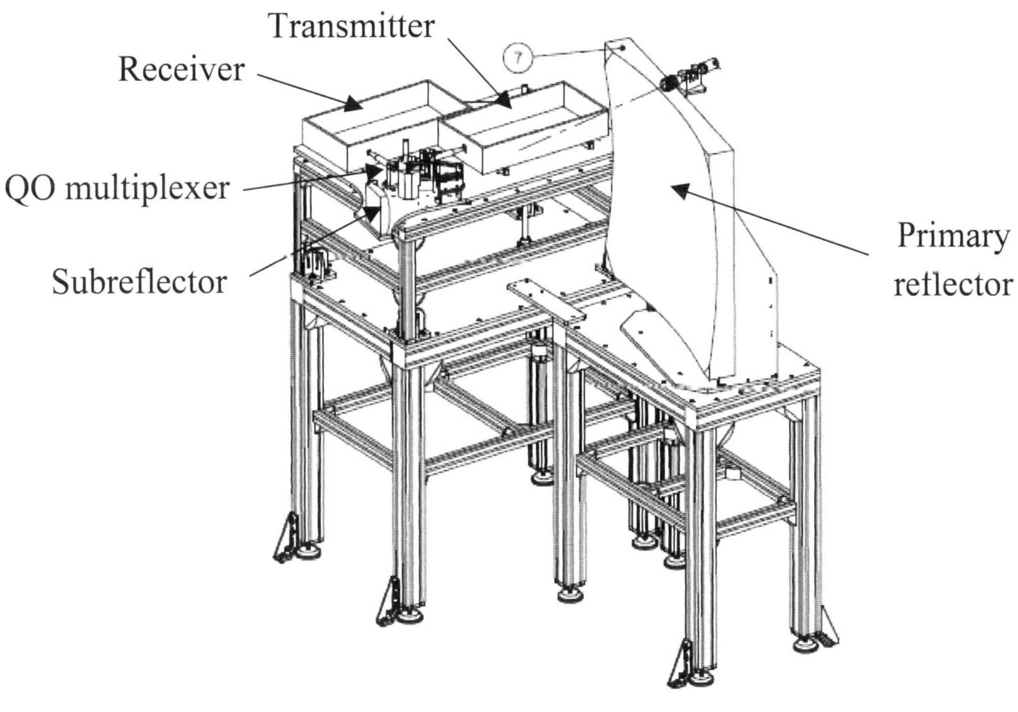

Figure1: The GRaCE hardware indicating the major elements.

The radar is designed to have the following parameters:

Parameter	Ground Radar	Space Radar	Comment: Ground/Space
Transmit frequency	200 GHz	160 – 260 GHz	G: Fixed, but can be set at design stage S: Fixed by EIK within ± 1 GHz
Transmit power	> 100 mW	100 W	G: Single high power Schottky multiplier
Transmitter Technology	Schottky	EIK	**G, S**: High power Schottky frequency multiplier **S:** Extra EIK linear amplifier required
Transmit polarisation	Circular	Circular	From high performance quasi-optical Tx/Rx diplexing
Receive polarisation			
Main antenna diameter	1 m	≥2 m	**G**: Set by project budget **S**: Set by satellite payload and budget
Beam divergence	≈ 0.1°	≈ 0.05°	**G, S**: Set by main antenna diameter
Antenna Gain	64 dBi	70 dBi	**G, S**: Set by main antenna diameter
Antenna sidelobe level	<40 dB	<50 dB	**S**: Very low sidelobes required to suppress ground reflections off-axis
Range resolution	50 to 500 m	500 m	**G:** Selectable in software, trade off with sensitivity **S**: Similar to CloudSat & EarthCARE
Pulse length / coding	20 ns to 1 μs	3.3 μs	Determines range resolution
Maximum instrument range	12 km	500 km	**G**: Few clouds above 11 km at UK latitudes **S**: Satellite altitude ~400 km, atmosphere top @ 380 km
Receiver Noise Figure	6 dB	6 dB	Set by receive mixer and optics / radome losses
Receiver Technology	Schottky	Schottky	**G, S**: Sub-harmonically pumped Schottky diode mixer, potentially with LNA
Sensitivity	-26 dBZ	-20 dBZ	G: At 1 km range, single 400 ns pulse **S**: At 400 km range, single shot

This paper covers the development of the 200 GHz multiplexing antenna for GRaCE.

Unusually, the multiplexer does not rely upon gyrotropic materials (ferrites) as would typically be used for linear polarisation. Instead it follows an approach used in the JAXA 94 GHz radar in the ESA/JAXA EarthCARE mission [2], involving transmission and reflection of circularly polarised beams. In that case, one of two corrugated horns attached to the sources transmits a beam to a mechanically switchable mirror (coloured brown in left hand part of Figure 2) and is reflected to a refocusing mirror (coloured purple) following which it passes through an analysing polarizing grid (coloured gold).

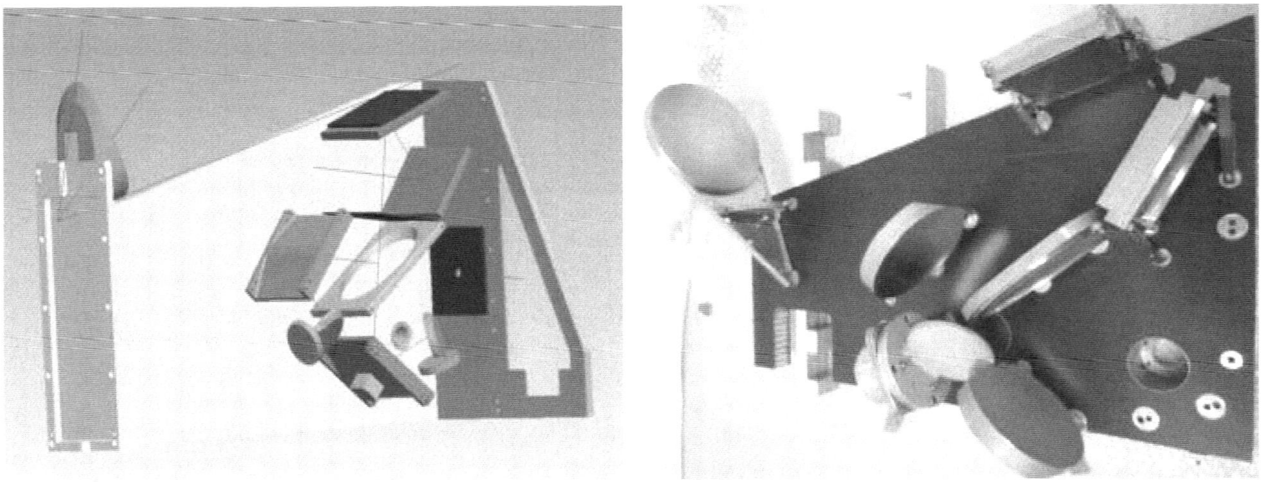

Figure 2: Design and realisation of the 94 GHz quasi-optical multiplexer produced by TK for the Cloud Profiling Radar on EarthCARE

The beam then passes to an Inatani-type Martin-Puplett Diplexer [3] which acts as the circular polarizer (coloured blue-green). The radiation then passes to the sub reflector (coloured green) and up to the main reflector, not shown, for transmission.

The returning beam, reflected by hydrometeors, follows a reversed path, being diverted sideways at the analysing grid (coloured gold), and a refocusing mirror (coloured blue). A second switchable mirror directs the beam to one of two corrugated horns feeding the low noise amplifier receivers. The switching mirrors provide transmit and receive redundancy in the space mission: they are not required for GRaCE.

The scheme used for EarthCARE has been adapted to operate at 200 GHz for GRaCE. Mechanical design was performed in PTC CREO – Figure 3.

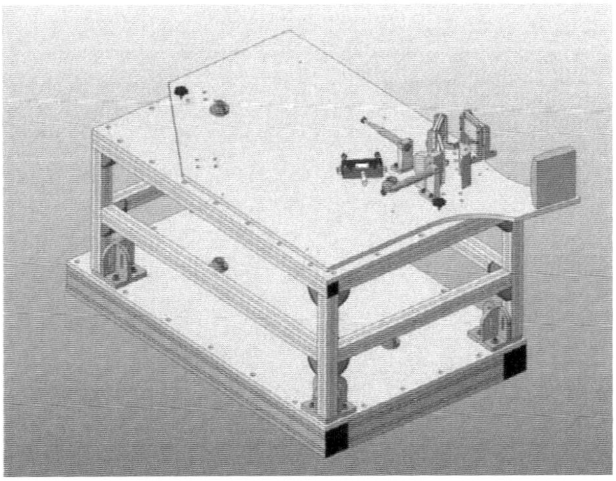

Figure 3: Design of the GRaCE optics bench, with the 200 GHz quasi-optical multiplexer and sub-reflector.

Figure 4 shows images of the constructed quasi-optical muliplexer, with a pair of ultra-Gaussian feedhorns, and the system under test at RAL.

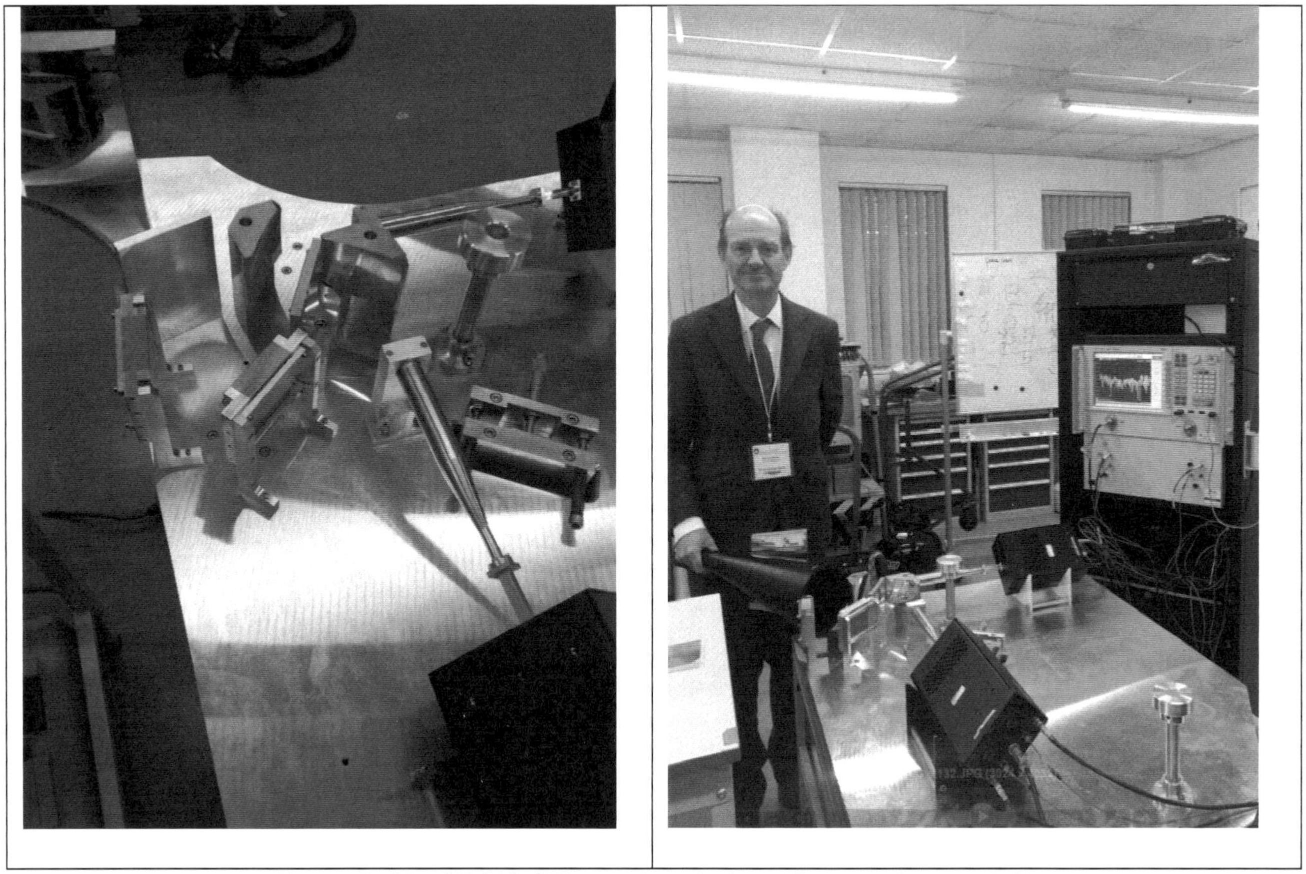

Figure 4: GRaCE 200 GHz quasi-optical multiplexer close up (left) and under VNA test at RAL with Dr Richard Wylde of TK Ltd. (right).

Insertion loss and isolation measurements were performed using a vector network analyser and frequency extender modules, Figure 4 right. At 200 GHz, the single pass insertion loss of the network is at the 0.6 dB level, and transmit/receive isolation at a very useful 60 dB.

Testing of the whole instrument at STFC's Chilbolton Observatory is to start shortly.

[1] ceoi.ac.uk

[2] www.esa.int/Applications/Observing_the_Earth/The_Living_Planet_Programme/Earth_Explorers/EarthCARE/Satellite

[3] smiles.tksc.jaxa.jp/old/document/article1999/inatani_thz1999_fsp.pdf

1 to 8 beam distributor at 4.7 THz for GUSTO

B. Mirzaei[1,2], M. Finkel[1], J. R. Silva[1], W. Laauwen[1], C. Groppi[3], A. Khalatpour[4], Q. Hu[4], A. Young[5], C. Walker[5], and J. R. Gao[1,2]

NASA GUSTO (Galactic/Extragalactic ULDB Spectroscopic Terahertz Observatory), equipped with three 8-pixel heterodyne detection channels, will perform the largest, single-flight mapping of three important interstellar lines of nitrogen [NII], carbon [CII], and oxygen [OI]. GUSTO will provide comprehensive understanding of the life-cycle of the interstellar medium in Milky Way and Large Magellanic Cloud (LMC). The local oscillators (LO) coupling to the state of the art mixers applied in GUSTO is quite crucial for all three channels and more challenging at [OI] frequency of 4.7 THz, where frequency multiplied sources are not available. For the latter frequency applying a single quantum cascade laser (QCL) beam as the LO multiplexed to 8 by a phase grating is shown to be the most practical and efficient way [1].

In GUSTO a 4.7 THz QCL beam will be coupled to the mixer array through a warm optical unit, allowing for spatially filtering, steering and multiplexing. There the QCL amplitude stabilization and spectral monitoring will take place too. The geometrical limits in that unit demand a certain combination of the input beam size and the angular separation of the output beams, which can only be addressed by a grating with an asymmetric profile. The design of such a grating and the influences of its tolerances on the overall coupling were presented in the previous ISSTT. Here we report the grating manufacturing and its experimental qualifications to give a full study of the instrument driven grating.

A CNC (Computer Numerical Control) micro-milling machine at Arizona State University (ASU) is used to transform the designed surface profile onto an aluminum plate, where the post-machining profile measurements using a 3D microscope were done too. The achieved surfaces were evaluated at SRON/TUDelft, where a deviation less than 1 μm in height from the design is found.

We characterized one of the manufactured gratings using an engineering GUSTO QCL (from MIT) as the input source, which emits a single-mode beam at 4.715 THz with an output power of about 5 mW operating at 60 K. We focused on the angular separation, power distribution among the output beams, and the grating efficiency (the ratio between the total power of the wanted diffracted beams and the input). We found the angular distribution agreeing to the simulation within an experimental accuracy of ≤ 0.5°. We also found the efficiency to be 70%, and the non-uniformity (the difference between the maximum and minimum powers divided by the average power from all 8 beams) to be 13%, both agreeing well to what is expected from the simulation.

It is the first time that such a complete characterization of a THz phase grating is being reported. Figure 1 shows the output beams measured at a distance of 34 cm from the grating, together with the input beam, mapped on the same plane when the grating is replaced with a flat mirror.

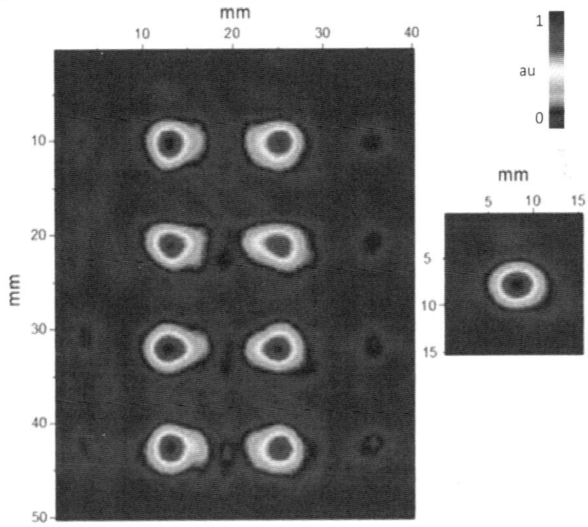

Fig. 1. The measured 8-beam pattern (left) and the incident beam on the same plane when the grating is replaced with a flat mirror (right). The color indicates the intensity.

In conclusion, we designed, manufactured, measured and confirmed a phase grating as the 1 to 8 beam multiplexer for 4.7 THz channel of GUSTO.

[1] SRON Netherlands Institute for Space Research, the Netherlands
[2] Faculty of Applied Science, Delft University of Technology, the Netherlands
[3] School of Earth and Space Exploration, Arizona State University, USA
[4] Department of Electrical Engineering and Computer Science, Massachusetts Institute of Technology, USA.
[5] Steward Observatory, University of Arizona, USA.

REFERENCES

[1] B. Mirzaei, et al, "8-beam local oscillator array at 4.7 THz generated by a phase grating and a quantum cascade laser," Opt. Express 25, 29587-29596 (2017).

NOTES:

Design, Fabrication and Characterization of Waveguide to Substrate Transition Based on Unilateral Substrateless Finline Structures

C. Lopez[1], V. Desmaris[1], D. Meledin[1], A. Pavolotsky[1] and V. Belitsky[1]

Waveguide-to-substrate transitions play a fundamental role in any THz system since active components are realized using thin film technology. The main requirements of such transitions are good impedance match and ease of fabrication.

Due to its large fractional bandwidths, waveguide E-probes have been widely adopted in the field [1,2]. Nevertheless, its performance is widely affected by its position inside the waveguide. Unilateral finline structures have been proposed [3] as an alternative solution. However, this approach results in problems related to impedance matching, which affects its performance over large operational bandwidths.

We present a novel waveguide to substrate transition, for prospective use in broadband mixer. The high impedance of the full-height waveguide is matched to the slotline using a unilateral substrateless finline structure and a 2-section slotline Chebyshev transformer.

The transition is designed of a silicon substrate covered by a superconducting niobium thin layer. In order to reduce the overall insertion loss and achieve the correct matching, the substrate encompass by the fins is fully removed. A gold layer located on top of the superconducting film provides grounding for the fins and aids the mounting process in the slit-block waveguide mount. Furthermore, the waveguide width is reduced in steps to reach a subcritical square cavity.

For the sake of comparison between simulations and measurements, a back-to-back arrangement was designed for the 211-373 GHz frequency band. Simulation results for this structure showed a fractional bandwidth of 55% with a return loss better than 15 dB. Moreover, a second design was developed for room temperature operation, where the superconducting film has been replace by gold.

The fabricated cryogenic transitions were characterized at 4K using a cryostat. The test setup also included a Rohde & Schwarz ZNA43 VNA with frequency extension modules operating at room temperature.

At the Conference, simulations and measured results for both, cryogenic and room temperature transitions, will be presented.

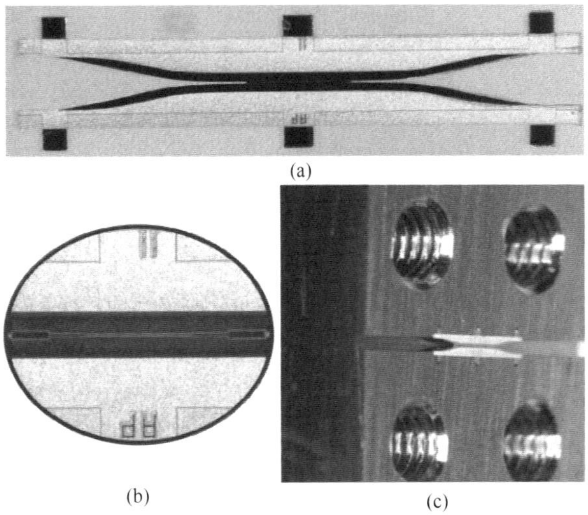

Fig. 1. (a) Photograph of the fabricated back-to-back transition. (b) Detail of the central slotline Chebyshev transformer. (c) Device mounted in the split-block waveguide. Silicon tips facilitate the mounting process and aids the handling of devices.

REFERENCES

[1] Risacher, C., Vassilev, V., Pavolotsky, A., & Belitsky, V. "Waveguide-to-microstrip transition with integrated bias-T.", IEEE Microwave and Wireless Components Letters, vol 13, pp. 262-264, 2003.

[2] J. W. Kooi, G. Chattopadhyay, S. Withington, F. Rice, J. Zmuidzinas, C. Walker, & G. Yassin, "A full-height waveguide to thin-film microstrip transition with exceptional RF bandwidth and coupling efficiency," International Journal of Infrared and Millimeter Waves, pp.261-284, 2003.

[3] Tan, Boon Kok, and Ghassan Yassin, "Unilateral finline transition at THz frequencies," Twentieth International Symposium on Space Terahertz Technology", 2009

[1] Group for Advanced Receiver Development, Chalmers University of Technology, SE412 96, Gothenburg, Sweden.

Spline-Profile Diagonal Horn Transmitter at 104 GHz Suitable for LLAMA Observatory Holography Measurements

Daniele Ronso Lima[1,2], Rocio Molina[3], Catalina Medina[3], Danilo C. Zanella[1], Jacques R. D. Lèpine[1], Andrey M. Barychev[2].

The Large Latin American Millimeter Array (LLAMA) is a joint scientific and technological project of Argentina and Brazil whose goal is to install and to operate a radio observatory in the Argentine Andes, at 4800m of altitude. LLAMA observatory is a single dish 12m parabolic antenna made by Vertex company, as APEX telescope.

Currently, we are designing the holography facilities and devices to guarantee the functionality of the antenna. Holography measurements requires a narrow bandwidth and a strong phase uniformity to ensure the accuracy of the manufacture specifications in a parabolic antenna surface. The antenna as delivered may have the panels mis-adjusted such that the total error is as large as 100μm; Final adjustments should be done by the LLAMA team.

The basic specification is that the final surface should be accurate to 20μm, computed as the RSS of various contributions. Using the holography recommendations from ALMA [1], it is also required to allow for 10μm of RMS error in panel setting knowledge (and 2μm of error in setting the panels to the desired places.

This paper is about the design and characterization of the transmitter horn's holography system. The transmitter's beam width is determined by the need to illuminate the 12 m antenna with reasonable uniformity. Variations in phase, if uncorrected, lead directly to errors in the reflector surface deviations being measured. It is expected that the horn can be made to produce a far-field spherical wave to the required tolerance within the part of the beam being used.

Corrugated horns have traditionally been the preferred choice with the best performance but are expansive and impractical to manufacturing when the frequency is increased. We are using the Gibson Model [2], a split block diagonal horn optimized for 104GHz [3]. The performance of this antenna is less than a corrugated horn, but optimizing the depth profile, the performance can be improved significantly using only a uniquely simple milling machine.

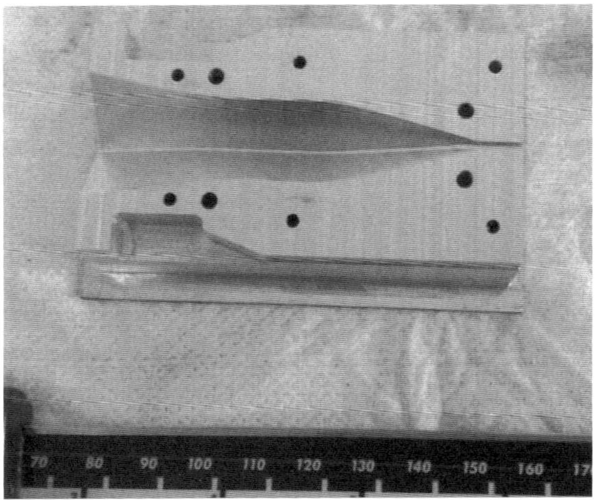

Fig. 1. Mechanical model of the Spline-Profile Diagonal Horn Antenna.

In this paper we will show the model, the design, the measurements procedures and the results of the characterization. Furthermore, we will compare the results with the previous specifications of the horn transmitter and discuss it.

References

[1] L. D'Addario, "ALMA Holography System Overview" ALMA Holography CDR documents. NRAO Tucson, 2000-October-10.

[2] H. J. Dibson, B. Thomas, L. Rolo, M. C. Wiedner, A. E. Maestrini, P de Maagt, "A Novel Spline-Profile Diagonal Horn Suitable for Integration Into THz Split-Block Components" IEEE Transactions on Terahertz Science and Technology, pp 657-663, vol 7, Issue: 6 , Nov. 2017.

[3] J. K. Johansson and N. K. Whyborn, "Antennas for sub-millimetre wave receivers," in Proc. 2nd Int. Symp. Space THz Technol. (ISSTT 1991), Pasadena, CA, USA, Feb. 26–28, 1991, pp. 63–81.

[1] University of São Paulo, São Paulo, Institute of Astronomy, Gephysics and atmospherical sciences. National Institute of Standards and Technology, Boulder, CO 80305 USA.

[2] Kapteyn Astronomical Institute, University of Groningen, Postbus 800, 9700 AV Groningen, The Netherlands e-mail: r.hesper@astro.rug.nl

[3] Astronomy Department, Universidad do Chile, Camino El Observatorio 1515, Santiago, Chile.

Development of THz Superconducting HEB Receiver Systems for Balloons, Aircraft, SmallSats and Future Large Missions

Jonathan Kawamura[1], Daniel Cunnane[1], Boris Karasik[1], Bruce Bumble[1], Darren Hayton[1], Imran Mehdi[1], Goutam Chattopadhyay[1], Maria Alonso[1], Jose Siles[1], Christopher Curwen[1], and Benjamin Williams[2]

We describe the on-going development of several classes of receiver systems based on superconducting hot-electron bolometer mixers. We target four general platforms, listed in time order, from immediate to long-term: high-altitude balloons (ASTHROS), aircraft (e.g., SOFIA), SmallSat and a future flagship mission (OST). Each platform has its own advantages and disadvantages in terms of resources and capability, and these have guided our multi-pronged effort to develop systems that can be suited for these opportunities.

For the near term, we are developing a receiver system for the ASTHROS mission. A low-power cooler capable of reaching 4 K operating temperature that consumes only 200 W allows us to use NbN-based waveguide mixers of modest array size (~8). These mixers are pumped by an array of frequency-multiplier chains image to the mixer array. The small number of pixels is compensated by the long duration of the flight, which is not limited by boil-off of helium as in other cryogenic balloon experiments.

An aircraft such as SOFIA can provide very generous power resources, and an instrument designed for it can be made very aggressive. We have proposed an instrument called SHASTA, which ultimately would have a 128 pixel array targeting the 1.9-2.1 THz band. As part of this effort, we are developing a modular NbN-based 16-pixel waveguide mixer arrays that could be tiled together in a future instrument. This system would use local oscillator arrays that permit horn-to-horn imaging between the mixer and LO arrays. To permit observations at higher frequencies target other spectroscopic lines, our development effort includes a 16-pixel 5 THz waveguide mixer array that is pumped by a single quantum-cascade laser source.

A SmallSat experiment is likely to be only feasible if a 20 K-class cooler is used. We are developing MgB2-based mixers, both in quasi-optical and waveguide configurations. Such a system can be pumped by FMCs for frequencies < 3 THz or by QCLs > 2 THz. We are currently developing a brassboard utilizing a commercial 20-K class cooler.

Finally, with a look into the future, a concept for a highly integrated mixer focal plane array is being developed. This would involve coupling low profile micro-machined silicon lenslet arrays fed by a planar array of balanced HEB mixers. The rf and lo port would be opposite sides of a flat mixer package, with LNAs integrated in proximity of the mixer and the IF signals fed out laterally.

We will present status on the development of all four systems.

[1] Jet Propulsion Laboratory, California Institute of Technology, Pasadena, California, 91109 USA.

[2] Department of Electrical Engineering, University of California at Los Angeles, Los Angeles, California, 90095 USA

NOTES:

Photo-Induced Coded-Aperture Terahertz Imaging using Mesa-Array Structures for Approaching Subwavelength Resolution

Yijing Deng*, Yu Shi, Jun Ren, Patrick Fay, and Lei Liu
Department of Electrical Engineering, University of Notre Dame, Notre Dame, IN 46556, USA
*Contact: ydeng2@nd.edu

Abstract— **We present a novel and unique solution for approaching subwavelength-resolution terahertz (THz) photo-induced coded-aperture imaging (PI-CAI) by utilizing micromachined mesa-array structures. In this approach, photo-induced free carriers are confined inside each mesa, facilitating much more refined photopatterns for achieving subwavelength spatial resolution. Mesa-array structures with a unit cell dimension of 105 µm × 105 µm, and a trench width of 7 µm were fabricated and tested. An average of ~11 dB modulation depth was obtained in the frequency range of 740-750 GHz under a light intensity of only 4 W/cm^2. Initial imaging experiments have been performed, and the results meet the expectations quite well.**

I. Introduction

In recent years, terahertz (THz) frequency regime has attracted increasing interest due to its emerging applications in medical imaging [1], security screening [2], and radio astronomy [3]. THz imaging has been extensively explored for the above applications utilizing array imagers (e.g., focal plane arrays) [4] and single-element imagers [5] (e.g., scanning probe microscopy). However, these prior approaches either require large-scale arrays of detectors or raster mechanical scanning, leading to increased system complexity and limited imaging performance (e.g., speed and resolution).

To overcome the above problems, coded-aperture imaging (CAI) technique [6] which is promising for realizing both system simplicity (e.g., a single-element detector) and high performance (e.g., high SNR and speed) has been proposed and demonstrated. Schottky diodes [7] and graphene modulators [8] have been reported to realize CAI aperture masks (for spatial modulation of THz waves). However, restricted by biasing circuitry, pixel sizes cannot be further scaled down, resulting in low imaging resolution and limited mask reconfigurability. An alternative approach is to employ photo-induced (PI) free carriers in semiconductors to generate reconfigurable CAI masks without the need for additional, complicated circuit prepatterning or device fabrication/integration [9]. However, due to lateral diffusion of PI free carriers, the achievable spatial resolution is limited, restricting its applications for imaging requiring higher resolution.

In this paper, we report a novel and unique solution for approaching subwavelength-resolution THz photo-induced coded-aperture imaging (PI-CAI) by utilizing micromachined mesa-array structures. The proposed structures utilize a two-dimensional array of semi-isolated mesas with lateral dimension much smaller than the wavelength. By confining PI free carriers inside each mesa, much more refined photopatterns and consequently much smaller mask pixels can be achieved to potentially demonstrate subwavelength imaging resolution. Si mesa-array structures with a unit cell dimension of 105 µm × 105 µm, and a trench width of 7 µm were fabricated and tested. An average modulation depth of ~ 11 dB was obtained in the frequency range of 740-750 GHz under a light intensity of ~ 4 W/cm^2. Initial imaging experiments were performed at 740 GHz, and the results meet the expectations quite well. More advanced THz imaging will soon be performed for demonstrating subwavelength resolution using the proposed mesa-array approach.

Fig. 1. Schematic drawing of mesa-array structures and zoom-in detail. The red dotted circle indicates a w_i-wide small connecting island at the corners of adjacent mesas (with a mesa width of w_m, and a trench width of w_t).

II. Optical Terahertz Spatial Modulation of Mesa-Array Structures

It has been demonstrated from our previous work that the THz transmittance of a semiconductor can be spatially modulated by changing the incident light intensity and the illuminated light pattern [10]. This spatially-resolved optical modulation (SROM) method could be applied to realize tunable and reconfigurable THz devices such as modulators, variable attenuators, and filters. However, due to lateral diffusion in naturally-existing semiconductors, the achievable spatial resolution of the resulted photopatterns from SROM is limited, restricting the implementation of tunable/reconfigurable devices in the THz regime [11].

To overcome this problem, we propose a novel optical approach to improve the spatial resolution based on micromachined mesa-array structures. As shown in Fig. 1, this structure consists of a two-dimensional array of subwavelength (e.g., w_m = 10 µm, much smaller than the lateral diffusion

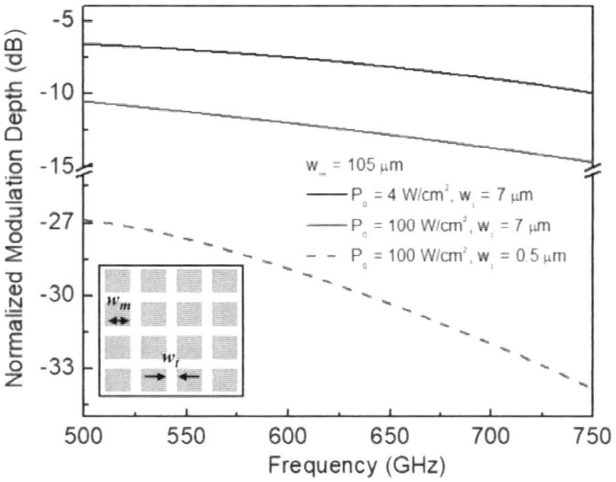

(a)

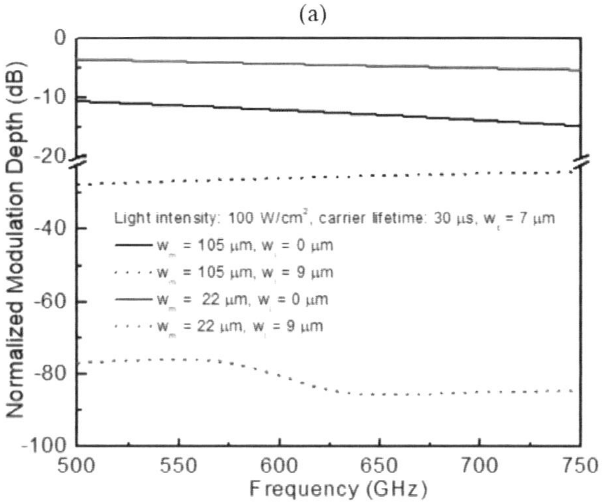

(b)

Fig. 2. (a) Optical spatial modulation depth of Si mesa-array structures with w_m = 105 µm, h = 50 µm, w_i = 0 (without connecting islands), and different w_t (i.e., 7 µm, and 0.5 µm). (b) The modulation depth comparisons for mesa-array structures (w_t = 7 µm, h = 50 µm, w_m = 105 µm or 22 µm) with (w_i = 9 µm) and without (w_i = 0) connecting islands. Different light intensities (4 W/cm² and 100 W/cm²) were employed, assuming a 30-µs carrier lifetime.

length) semiconductor mesas, resulting in photo-induced free carriers being confined inside each mesa. Consequently, reconfigurable PI-CAI masks with much smaller pixel sizes can be generated (since "blurring" of patterns caused by lateral diffusion is eliminated) to potentially achieve THz imaging with subwavelength resolution.

To validate the proposed approach, the optical modulation properties of Si mesa-array structures were first evaluated through HFSS simulations, assuming a free carrier lifetime of 30 µs. Fig. 2(a) shows the modulation depth of mesa arrays (w_m = 105 µm, substrate thickness h = 50 µm) with fully isolated mesas (see the inset in Fig. 2(a)). It can be seen that the modulation depth only increases by ~5 dB at 625 GHz as the light intensity P_0 increases from 4 W/cm² to 100 W/cm², for mesa arrays with trench width w_t = 7 µm. Although the modulation depth can be further increased by employing smaller w_t (e.g., ~30 dB at 625 GHz with w_t = 0.5 µm), the fabrication process for such structure could be extremely challenging. An alternative solution is to add small connecting islands (indicated in the red dotted circle in Fig. 1 inset) among adjacent mesas to keep the largest gap dimensions much smaller than the wavelength. Therefore, THz wave propagation through higher order modes can be effectively suppressed, leading to a higher achievable modulation depth (under the same light intensity). Fig. 2(b) shows the modulation depth improvement by utilizing the modified mesa array designs. It can be seen that by adding the connecting islands (w_i = 9 µm), the modulation depth for modified mesa arrays with w_m = 105 µm is improved by ~13 dB at 625 GHz under a light intensity of 100 W/cm². A larger increase in the modulation depth can be observed for mesa arrays with smaller w_m. Assuming w_m = 22 µm, the modulation depth could be increased by ~80 dB at 625 GHz under the same light intensity (see red lines in Fig. 2(b)). This is attributed to a larger difference in the gap dimensions before and after adding the small connection islands.

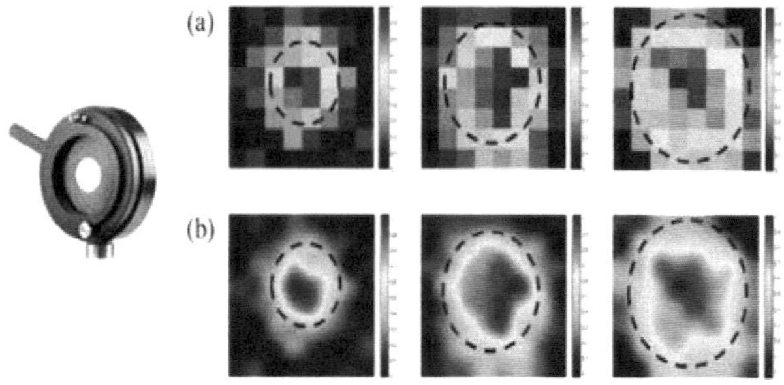

Fig. 3. THz imaging (8×8 pixels) of an iris aperture at 740 GHz using Si mesa-array structures: (a) reconstructed images of iris aperture with different sizes, (b) median filtered images for clarity. Black dashed circles highlight the dimensions of the apertures. Image size: ~12 mm × 9 mm.

III. Initial Imaging Experiments and Discussion

A Si mesa-array structure (w_m = 105 µm, w_t = 7 µm, w_i = 9 µm, h = 50 µm) was fabricated through deep reactive-ion etching (DRIE) process. An etching rate of ~5 µm/min was obtained using a combination of SF_6/C_4F_8 for etching and passivation.

The Si mesa-array structure was then tested using a WR-1.5 vector network analyzer (VNA) as the source and the detector. An optical modulation depth of ~11 dB was obtained under a light intensity of only 4 W/cm² in the frequency range of 740-750 GHz as expected. Once again, ~84 dB could be potentially achieved using 22-µm mesas under 100 W/cm² light intensity as predicted from simulation.

Initial THz imaging experiments were also performed at 740 GHz using the proposed Si mesa-array structures. On the basis of Hadamard coding, optical mask patterns can be generated through a digital mirror device (DMD) chip in a digital light processing (DLP) projector. Fig. 3 shows the reconstructed images (8×8 pixels, with an image size of ~12 mm × 9 mm) of an iris aperture with different sizes. It can be seen that the correpsonding outlines of the iris aperture (as shown by the black dashed circles) are clearly shown in the reconstructed images, indicating the validity of the PI-CAI using mesa-array structures. More advanced imaging experiments will be soon performed using more pixels (e.g., 16 × 16 pixels) with smaller pixel sizes to demonstrate THz PI-CAI with potentially subwavelength spatial resolution.

Conclusions And Future Works

A novel approach for realizing THz PI-CAI with potentially subwavelength spatial resolution is reported using micromachined mesa-array structures. The optical modulation properties of the proposed mesa arrays were first evaluated through full-wave HFSS simulations. A mesa-array structure was then fabricated and tested, and initial imaging results meet the expectations quite well. Imaging using more pixels will be soon performed to demonstrate potentially subwavelength THz spatial resolution. The proposed PI-CAI technique using mesa-array structures provides a novel approach for realizing THz imaging with subwavelength resolution.

ACKNOWLEDGMENT

This work is partially supported by the National Science Foundation (NSF) under the grant of ECCS-1711631 and ECCS-1711052, and a subcontract from the Harvard-Smithsonian Center for Astrophysics under grant PTX-Smithsonian 17-SUBC-400SV787007. The authors would like to thank the support from Center for Nano Science and Technology (NDnano) and Advanced Diagnostics & Therapeutics (AD&T) at the University of Notre Dame.

REFERENCES

[1] P. H. Siegel, "Terahertz technology in biology and medicine," *IEEE Trans. Microw. Theory Tech.*, vol. 56, no. 10, pp. 2438-2447, 2004.

[2] A. W. M. Lee, B. S. Williams, S. Kumar, Qing Hu, and J. L. Reno, "Real-time imaging using a 4.3-THz quantum cascade laser and a 320×240 microbolometer focal-plane array," *IEEE Photon. Technol. Lett.*, vol. 18, no. 13, pp. 1415-1417, 2006.

[3] D. Meledin et al., "A 1.3-THz balanced waveguide HEB mixer for the *APEX* telescope," *IEEE Trans. Microw. Theory Tech.*, vol. 57, no. 1, pp. 89-98, 2009.

[4] L. Liu, H. Xu, A. W. Lichtenberger, and R. M. Weikle II, "Integrated 585 GHz hot- electron mixer focal-plane arrays based on annular-slot antennas for imaging applications," *IEEE Trans. Microwave Theory and Tech.*, vol. 58, no. 7, pp. 504–506, 2010.

[5] D. Zimdars, "High speed terahertz reflection imaging," *Proc. SPIE*, vol. 5692, pp. 255-259, 2005.

[6] W. L. Chan, K. Charan, D. Takhar, K. F. Kelly, R. G. Baraniuk, and D. M. Mittleman, "A single-pixel terahertz imaging system based on compressed sensing," *Appl. Phys. Lett.*, vol. 93, no. 12105, 2008.

[7] Hawasli, S., Alijabarri, N. and Weikle II, R. M., "Schottky diode arrays for submillimeter-wave sideband generation," *Proc. Int. Conf. on Infrared, Millim., and THz Waves*, 2012.

[8] B. Sensale-Rodriguez, S. Rafique, R. Yan, M. Zhu, V. Protasenko, D. Jena, L. Liu, and H. G. Xing, "Terahertz imaging employing grapheme modulator array," *Opt. Exp.*, vol. 21, no. 2, pp. 2324–2330, 2013.

[9] L. Cheng and L. Liu, "Optical modulation of continuous terahertz waves towards cost-effective reconfigurable quasi-optical terahertz components," *Opt. Exp.*, vol. 21, no. 23, pp. 28657-28667, 2013.

[10] A. Kannegulla, M. I. B. Shams, L. Liu, and L.-J. Cheng, "Photo-induced spatial modulation of THz waves: opportunities and limitations," *Opt. Exp.*, vol. 23, no. 25, pp.32098-32112, 2015.

[11] Y. Deng, J. Ren, Y. Shi, Y.-Ch. Wang, L.-J. Cheng, P. Fay, and L. Liu, "Advanced photo-induced substrate-integrated waveguides using pillar-array structures for tunable and reconfigurable THz circuits," *Opt. Exp.*, vol. 28, pp. 7259-7273, 2020.

A 1.37 THz Waveguide-based 2 X 2 Beam Divider Fabricated by Two Microfabrication Technologies

Haotian Zhu[1], Jerome Valentin[1], Thibaut Vacelet[1], Sylwester Bargiel[2], Samuel Queste[2], Laurent Robert[2], Djaffar Belharet[2], Etienne Herth[3], Yan Delorme[1], and Martina Wiedner[1]

To improve the mapping speed, multi-pixel radio astronomy receivers are favored [1]. Observation speed will be increased by a factor of the pixel number of the receiver. Therefore, high science throughput can be produced.

Currently, there are only few heterodyne array receivers at THz. One of the challenges is the distribution of the local oscillator (LO) beam to all pixels. Often phase gratings are used to split a single LO beam into multiple beams. Here we propose a waveguide-based beam divider. In order to evaluate the performance of waveguide splitters, we use a single LO generated by a VDI Amplifier / Multiplier Chain. The LO beam is captured by a horn, transferred to waveguide and T-junctions split the signal first in two signals and then in four signals. These signals exit through circular horns into free space and are then overlayed onto the astronomical signal using a beam splitter [2]. The two signals are subsequently focused by lenses onto HEB mixers. We designed a waveguide splitter for 1.37 THz. The cross-sectional size of the WR0.65 waveguide is only 164 µm X 82 µm and is challenging to fabricate.

The beam divider contains two parts: feeding network and circular horn array [2]. The feeding network consists of one E-plane junction and two H-plane junctions. The feeding network has been fabricated using two technologies: a new technology, named Femtosecond laser assisted wet etching or 3D-laser microfabrication technology [3], and silicon etching using an inductively coupled plasma-deep reactive ion etching (ICP-DRIE) [4]. The 3D-laser microfabrication can fabricate highly accurate 3D -geometries based on fused silica by direct f-s laser writing and wet etching. The SEM pictures of the fabricated feeding networks by the two technologies are shown in Fig. 1. The total thickness of the feeding network is 400 µm, and the designed waveguide channel depth is 82 µm. A comparison of both 3D-laser and ICP-DRIE microfabrication is summarized in Table I. The feeding network fabricated by 3D-laser was plated with a 1.9-µm thick gold layer. The 3D-laser is easier to fabricate, but the bottom layer is rougher. On the other hand, the side wall of the feeding network manufactured by the ICP-DRIE is rougher. The estimated simulated waveguide insertion losses of the two types waveguide are 0.35 dB/mm and 0.31 dB/mm, as shown in the Table I. The length and the inner aperture diameter of the circular horn is 5 mm and 1.7 mm, respectively. The circular horn array is the only part fabricated by conventional mechanical machining. The 14.36-mm length waveguide in the feeding network contributes an additional insertion loss of 4.5-5 dB for each LO beam, according to the simulation. The first beam divider for demonstration has been assembled and will be characterized soon.

[1] LERMA, Observatoire de Paris, Sorbonne Université, Université PSL, CNRS, Paris, France.

[2] Institut FEMTO-ST, Université Bourgogne Franche-Comté UBFC, CNRS UMR 6174, 25030 Besançon, France.

[3] C2N, CNRS, Université Paris-Sud, Université Paris-Saclay, Palaiseau, France.

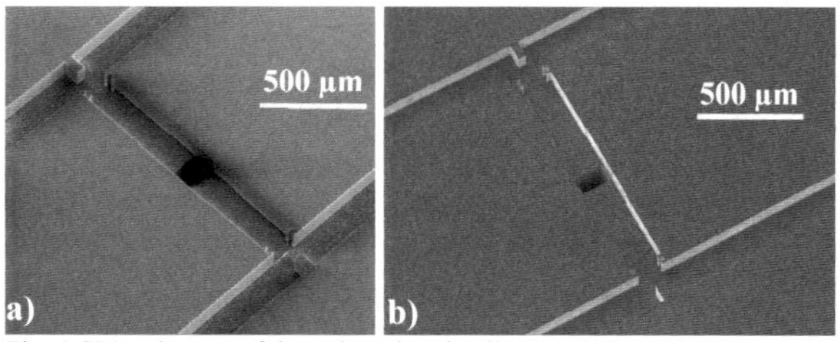

Fig. 1 SEM pictures of the T-junction details, (a) 3D-laser (b) ICP-DRIE

Table I. Comparison of the two manufacturing technologies investigated.

Technology	3D-laser	ICP-DRIE
Material	Fused silica	Silicon
Side angle	90 °	~89 °
Number of mask	Not needed	2
Mask misalignment	The issue does not exist	<2 µm
Bottom surface roughness	Rq ~367 nm*	Rq ~28 nm
Channel depth	84~88 µm	82~83 µm
Estimated waveguide insertion loss	0.35 dB/mm	0.31 dB/mm

*The value is measured after gold deposition.

Acknowledgement

This project is funded by the European research program Radionet-AETHRA and the Centre National d'Etudes Spatiales (CNES). This work was partly supported by the French RENATECH network and its FEMTO-ST technological facility. The authors would also like to thank Jean-Louis Leclercq from INL, for the gold deposition.

References

[1] U. Graf *et al.*, "Terahertz heterodyne array receivers for astronomy," *Journal of Infrared, Millimeter, and Terahertz Waves.*, vol. 36, no. 10, pp. 896-921, 2015.

[2] H. T. Zhu *et al.*, "A 2 X 2 beam divider for an array local oscillator at 1.37 THz," *Proc. 44th Int. Conf. Infrared Millim. THz Wave*, Sep., pp. 1-2, 2019.

[3] F. Sima *et al.*, "Three-dimensional femtosecond laser processing for lab-on-a-chip applications," Nanophotonics, vol. 7, no. 3, pp. 613–634, 2018.

[4] E. Herth *et al.*, "Fast ultra-deep silicon cavities: toward isotropically etched spherical silicon molds using an ICP-DRIE," *J. Vac. Sci. Technol. B*, vol. 37, no. 2, p. 021206, 2019.

NOTES:

Absorber Optimization Study for the Terahertz Intensity Mapper (TIM)

Rong Nie, Reinier M.J. Janssen, Charles M. Bradford, Jeffrey P. Filippini, and Steven Hailey-Dunsheath

***Abstract*—We discuss the design and optimization of the absorber for the long-wavelength arm of the Terahertz Intensity Mapper (TIM), a balloon-borne spectrometer employing Kinetic Inductance Detectors (KIDs). Electromagnetic simulations of our design indicate in-band absorption efficiency over $\sim 80\%$ in both linear polarization modes. By developing custom transmission line model and mode-matching calculations, we find the absorption efficiency is affected by the absorber's reactive part and overall shape. These insights into the operation of this design provide guidance for its optimization for low-resistance absorber materials.**

***Index Terms*—Electromagnetic simulations, Kinetic inductance detector, Mode matching method, Terahertz Intensity Mapper, Transmission line theory.**

I. Introduction

UNDERSTANDING the history of star formation throughout cosmic time would provide significant insights into the processes of galactic evolution. The far-infrared (FIR) wavelength band contains valuable information to address this question: half of the total energy output from the cosmic star formation has been absorbed by interstellar dust and re-emitted in the FIR [1][2]. Un-extincted FIR spectral lines (e.g., [CII], [NII], [OI]) can reveal the composition of the interstellar medium (ISM) and the abundance of star-forming clouds. Traditional single-object observations suffer from limitations of individual galaxy brightness, angular resolution, and survey speed. Line intensity mapping (LIM) is an emerging technique that maps the *integrated* emission of spectral lines from galaxies [3], which appear at different wavelengths depending on the redshifts of their sources. LIM thus enables study of the ISM and galaxy evolution over cosmic time without demanding high angular resolution [4].

The low noise, broad bandwidth, and large survey area required for FIR LIM demands an observing platform above the bulk of the Earth's atmosphere. The Terahertz Intensity Mapper (TIM) [5] is a balloon-borne FIR spectrometer designed to observe key spectral line tracers at the epoch of peak cosmic star formation. With an observing bandwidth of 240–420 μm, TIM will observe spectral lines from [CII] (158μm, visible $0.5< z <1.6$), [NII] (205μm, $0.2< z <1$), [OI] (63μm, $2.8< z <5.7$), and [OIII] (88μm, $1.7< z <3.8$), spanning the crucial gap in spectroscopic coverage between ALMA and JWST. TIM will make a pioneering demonstration of LIM by observing both *spectral* and *spatial* signals. In addition, TIM will: (i) detect $\sim$100 galaxies to constrain models of galaxy evolution; (ii) capture the star formation contribution of galaxies too faint to be detected individually through the measurement of the [CII] luminosity function across the peak of cosmic star formation; (iii) stack [CII] and [NII] lines on stellar mass-selected galaxies with spectroscopic redshifts, informing theory to relate this to the total star formation rate, star formation mode, metallicity, and specific star formation rate; and (iv) measure the specific star formation rate as a function of redshift by cross-correlation with Herschel/SPIRE and Spitzer/IRAC. Moreover, TIM will be a vital technological and scientific stepping stone to future orbital missions such as the Origins Space Telescope [6].

The promise of FIR line-intensity mapping brings with it substantial technical challenges. TIM's 2 m warm reflector will feed an $R \sim 400$ grating spectrometer, partitioned into long wavelength (LW: 317–420 μm) and short wavelength (SW: 240–317 μm) modules. To approach photon-noise-limited performance with high scalability, TIM employs feedhorn-coupled aluminum Kinetic Inductance Detector (KID) arrays operated at 250 mK [7]. These KIDs will employ a lithographically-patterned aluminum absorber coupled to a waveguide to absorb the incoming radiation.

In this paper we discuss electromagnetic simulation results and optimization studies for a TIM absorber design based upon that used in the MAKO camera [8]. This is a short version for ISSTT 2020; in full version of this parper (IEEE special issue), we will discuss transmission line theory and the mode-matching method to give greater intuition for our results.

R. Nie and J. P. Filippini are with the Department of Physics, University of Illinois at Urbana-Champaign, Urbana, IL 61801 USA.

R. M.J. Janssen and C. M. Bradford are with NASA Jet Propulsion Laboratory, Pasadena, CA 91109 USA, and with the Department of Astronomy, California Institute of Technology, Pasadena, CA 91125 USA.

S. Hailey-Dunsheath is with the Department of Astronomy, California Institute of Technology, Pasadena, CA 91125 USA.

Manuscript submitted June 1, 2020

II. Finite-Element Simulations

Inspired by the design developed for MAKO, the proposed TIM absorber (**Fig.** 1) is a 30 nm thick and 400 nm wide aluminum (Al) meandering line forming a quasi-mesh structure, patterned onto a silicon wafer with integrated back-short. The MAKO absorber was developed for the high sheet resistance, R_s, of titanium nitride (TiN) KIDs. A key challenge for TIM is thus to adapt this design to the much lower resistivity ($R_{s\text{TiN}}/R_{s\text{Al}} \sim 100$) of aluminum.

Building off an initial design [7], we carried out a program of simulations using the finite-element method (FEM) electromagnetic software ANSYS-HFSS. **Fig.** 1 shows the simulation setup. Radiation is guided to the quasi-mesh absorber by a 127 μm radius circular waveguide that flares out to a 225 μm

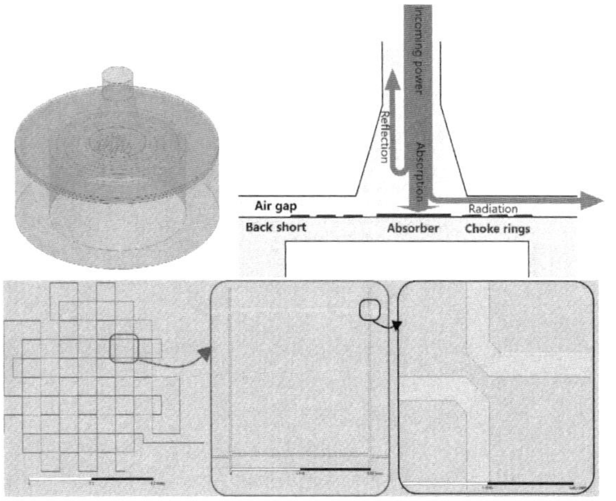

Fig. 1. *Top*: 3-D rendering in HFSS and side view cartoon of a single pixel. Photons propagate from the top port through a flared circular waveguide onto the aluminum meander, which is front-side illuminated and surrounded by three optical choke rings. A back-short layer is deposited on the other side of the silicon wafer. *Bottom*: Top view of the meander geometry, with insets highlighting mesh intersections. The various segments of meander line come close enough to one another at the corners to create capacitive shorts at the optical frequencies. The *Bottom Middle* panel is effectively a unit cell of the entire absorber. Throughout this paper, x and y refer to the vertical and horizontal axes of the lower pane.

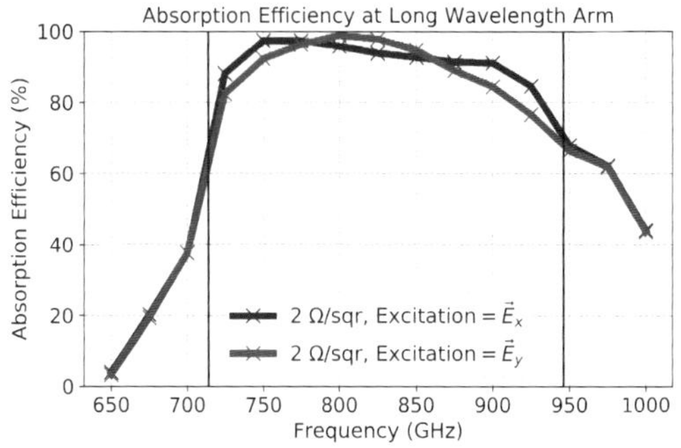

Fig. 2. Simulation results for the LW optimized design: air gap distance = 25μm, back-short distance = 27μm, line width = 0.4μm, the waveguide is excited by a single circular TE_{11} mode, and the absorber is surrounded by three choke rings. Vertical lines delimit the desired bandwidth (714-946 GHz).

radius near the absorber. The waveguide and absorber are separated by a short air gap. The absorber is located on a Si wafer with a metalized backside, which functions as a $\lambda/4$ back-short. In addition, the front-side illuminated absorber is surrounded by three choke rings made from the same aluminum as the absorber. In the optimized design the air gap and back-short distances are 25 μm and 27 μm, respectively. Both the choke ring widths and the gaps between them are 58 μm.

Power is propagated down the waveguide in the TE_{11} mode, since this is the only mode carried by the waveguide over a large part of the frequency band; the lowest three cut-off frequencies for a circular waveguide with radius a=127μm are TE_{11}=691.6 GHz, TM_{01}=903.6 GHz and TE_{21}=1147.4 GHz. The incoming power from the waveguide is distributed to five different ports, each measured independently in HFSS: absorption by the antenna (P_{ab}), reflection back to the waveguide ($|S_{11}|^2$), absorption by the choke rings (P_{cr}), and radiation escaping the pixel through the air gap (P_{ag}) and the substrate (P_{sb}). The reflected power comes from S_{11} directly, while the absorbed and radiated powers are defined as $P_i = \int \mathrm{Re}(\mathbf{S} \cdot \mathbf{n}) dA_i$, where **S** is the Poynting vector, and **n** is the unit outward normal vector of each surface A_i. As a consistency check, we computed the sum of these five components to confirm that they match the input power to <0.1%. Further details are given in the appendix.

We optimized over the following parameters: (a) air gap and (b) back-short distances, (c) meandering line width, and (d) total absorber size. We also explored the effects of (e) including higher-order circular waveguide modes, (f) switching to a rectangular waveguide, (g) excluding the choke structures, and (h) changing the choke ring dimensions. **Fig.** 2 shows the absorption efficiency of the optimized design. Absorption at low frequencies is suppressed by the waveguide cut-off frequency, while the high frequencies are reduced by the back-short thickness. Since TIM's detectors should ideally be sensitive to incoming signals independent of polarization, we investigate antenna performance in both x and y polarizations (blue and red lines in **Fig.** 2). We also simulate a representative input signal polarized at an intermediate angle and with an arbitrary phase delay ($E_{in} = \vec{E_x} + e^{0.28\pi i}\vec{E_y}$); the result is similar to the simple superposition of polarizations, suggesting that the independent study of x and y-polarization performance shown here is adequate.

III. Conclusion

Using finite-element simulations, we have demonstrated a meandered Al KID absorber for TIM with high absorption efficiency (~80%) in both x and y-polarizations. The power loss through the wafer and air gap is low (~0.1%), indicating minimal optical cross-talk between pixels. Based upon the success of this design, fabrication and testing of the first TIM test devices is anticipated in fall 2020.

Though we have demonstrated a successful design, we have several promising directions for future optimization. The MAKO-style absorber discussed here has relatively few parameters to tune (mainly meander line width and spacing), and the degree of polarization of the antenna depends upon these in non-trivial ways. The present design also relies on a relatively small air gap and a number of narrow (~0.3 μm) gaps at the mesh vertices (**Fig.** 1, lower right); while these parameters are manageable, a design with larger gaps and a wide tolerance in aluminum sheet resistance would be appealing. We are currently pursuing additional design efforts around an alternate absorber design that may prove more robust to fabricate for both LW and SW bands. This effort will be the subject of a future publication.

Acknowledgment

The authors thank Chris McKenney and Theodore Reck for useful scientific discussions. TIM is supported by NASA under grant 80NSSC19K1242, issued through the Science Mission Directorate. R.M.J. Janssen is supported by an appointment to the NASA Postdoctoral Program at the NASA Jet Propulsion Laboratory, administered by Universities Space Research Association under contract with NASA.

REFERENCES

[1] M. G. Hauser and E. Dwek, "The cosmic infrared background: Measurements and implications," *Annual Review of Astron and Astrophys*, vol. 39, pp. 249–307, Jan. 2001, doi: 10.1146/annurev.astro.39.1.249.

[2] G. Lagache, J.-L. Puget, and H. Dole, "Dusty Infrared Galaxies: Sources of the Cosmic Infrared Background," *Annual Review of Astron and Astrophys*, vol. 43, no. 1, pp. 727–768, Sep. 2005, doi: 10.1146/annurev.astro.43.072103.150606.

[3] E. D. Kovetz, M. P. Viero, A. Lidz, L. Newburgh, M. Rahman, E. Switzer, M. Kamionkowski, J. Aguirre, M. Alvarez, J. Bock, J. R. Bond, G. Bower, C. M. Bradford, P. C. Breysse, P. Bull, T.-C. Chang, Y.-T. Cheng, D. Chung, K. Cleary, A. Corray, A. Crites, R. Croft, O. Doré, M. Eastwood, A. Ferrara, J. Fonseca, D. Jacobs, G. K. Keating, G. Lagache, G. Lakhlani, A. Liu, K. Moodley, N. Murray, A. Pénin, G. Popping, A. Pullen, D. Reichers, S. Saito, B. Saliwanchik, M. Santos, R. Somerville, G. Stacey, G. Stein, F. Villaescusa-Navarro, E. Visbal, A. a. Weltman, L. Wolz, and M. Zemcov, "Line-intensity mapping: 2017 status report," *arXiv e-prints*, p. arXiv:1709.09066, Sep. 2017. [Online]. Available: https://arxiv.org/abs/1709.09066

[4] E. Kovetz, P. C. Breysse, A. Lidz, J. Bock, C. M. Bradford, T.-C. Chang, S. Foreman, H. Padmanabhan, A. Pullen, D. Riechers, M. B. Silva, and E. Switzer, "Astrophysics and Cosmology with Line-Intensity Mapping," *Bulletin of the AAS*, vol. 51, no. 3, p. 101, May 2019. [Online]. Available: https://arxiv.org/abs/1903.04496

[5] J. Vieira, J. Aguirre, C. M. Bradford, J. Filippini, C. Groppi, D. Marrone, M. Bethermin, T. Chang, M. Devlin, O. Dore, J. Fu, S. Hailey-Dunsheath, G. Holder, G. Keating, R. Keenan, E. Kovetz, G. Lagache, P. Mauskopf, D. Narayanan, G. Popping, E. Shirokoff, R. Somerville, I. Trumper, B. Uzgil, and J. Zmuidzinas, "The terahertz intensity mapper (tim): an imaging spectrometer for galaxy evolution studies at high-redshift," in *ISSTT2019*, Gothenburg, Sweden, Jun. 2019, pp. 208–215. [Online]. Available: https://www.nrao.edu/meetings/isstt/papers/2019/2019208215.pdf

[6] M. Meixner, A. Cooray, D. Leisawitz, J. Staguhn, L. Armus, C. Battersby, J. Bauer, E. Bergin, C. M. Bradford, K. Ennico-Smith, J. Fortney, T. Kataria, G. Melnick, S. Milam, D. Narayanan, D. Padgett, K. Pontoppidan, A. Pope, T. Roellig, K. Sandstrom, K. Stevenson, K. Su, J. Vieira, E. Wright, J. Zmuidzinas, K. Sheth, D. Benford, E. E. Mamajek, S. Neff, E. De Beck, M. Gerin, F. Helmich, I. Sakon, D. Scott, R. Vavrek, M. Wiedner, S. Carey, D. Burgarella, S. H. Moseley, E. Amatucci, R. C. Carter, M. DiPirro, C. Wu, B. Beaman, P. Beltran, J. Bolognese, D. Bradley, J. Corsetti, T. D'Asto, K. Denis, C. Derkacz, C. P. Earle, L. G. Fantano, D. Folta, B. Gavares, J. Generie, L. Hilliard, J. M. Howard, A. Jamil, T. Jamison, C. Lynch, G. Martins, S. Petro, D. Ramspacher, A. Rao, C. Sand in, E. Stoneking, S. Tompkins, and C. Webster, "Origins Space Telescope Mission Concept Study Report," *arXiv e-prints*, p. arXiv:1912.06213, Dec. 2019. [Online]. Available: https://arxiv.org/abs/1912.06213

[7] S. Hailey-Dunsheath, A. C. M. Barlis, J. E. Aguirre, C. M. Bradford, J. G. Redford, T. S. Billings, H. G. LeDuc, C. M. McKenney, and M. I. Hollister, "Development of Aluminum LEKIDs for Balloon-Borne Far-IR Spectroscopy," *Journal of Low Temperature Physics*, vol. 193, no. 5-6, pp. 968–975, Dec. 2018, doi: 10.1007/s10909-018-1927-y.

[8] C. M. McKenney, H. G. Leduc, L. J. Swenson, P. K. Day, B. H. Eom, and J. Zmuidzinas, "Design considerations for a background limited 350 micron pixel array using lumped element superconducting microresonators," in *Millimeter, Submillimeter, and Far-Infrared Detectors and Instrumentation for Astronomy VI*, W. S. Holland, Ed., vol. 8452, International Society for Optics and Photonics. SPIE, 2012, pp. 220 – 229, doi: 10.1117/12.925759.

Meissner Effect Transistor

*Siddhartha Sirsi and Christopher K. Walker

***Abstract*—The operation of conventional transistors is based on modulating the conductivity within a semiconductor by the application of an electric field. Meissner Effect Transistor (MET) has a complimentary device architecture within which the conductivity of a superconducting bridge is modulated by an applied magnetic field by way of the Meissner Effect. The upper cut-off operating frequency is dependent on Cooper pair relaxation time. Here we introduce the theory behind the operation of the MET, as well as initial proof-of-concept laboratory measurements.**

***Index Terms*—Cooper Pair, HEB, Meissner Effect**

I. Theory of Operation

Superconductors are made by forming Cooper pairs between electrons. The electrons in a Cooper pair have opposite (antiparallel) spins. Magnetic fields work to align the electron spins. Meissner and Ochsenfeld (1933) found that if a superconductor is cooled in a magnetic field to below the transition temperature T_c, then at the transition, the lines of induction B are pushed out. Likewise by imposing a magnetic field, B, on a superconductor it is possible to break the Cooper pairs in a controlled way, thereby modulating the conductivity of the superconductor.

The theory of operation for a Meissner Effect Transistor (MET) is analogous to a Field Effect Transistor (FET). A standard FET has 3 ports formed by metallic electrodes deposited on a semiconductor substrate.These ports are the gate (G), source (S), and drain (D). A time varying electric field applied to G modulates the conductivity in the underlying channel exponentially; i.e. small changes in the time varying gate voltage can cause large variations in the channel conductivity. In case of MET, the ``channel'' is a thin superconducting bridge and the gate electrode is replaced by an oscillating magnetic field, H_a that modulates the bridge's conductivity also in a nonlinear manner (Fig. 1). The greatest gain (or sensitivity) occurs when the bridge is magnetically or thermally biased near the point of going normal

A linear small signal model for an MET can be derived using the same approach as is used for an FET[1]. The small signal drain current i_D is a function of both the small signal drain

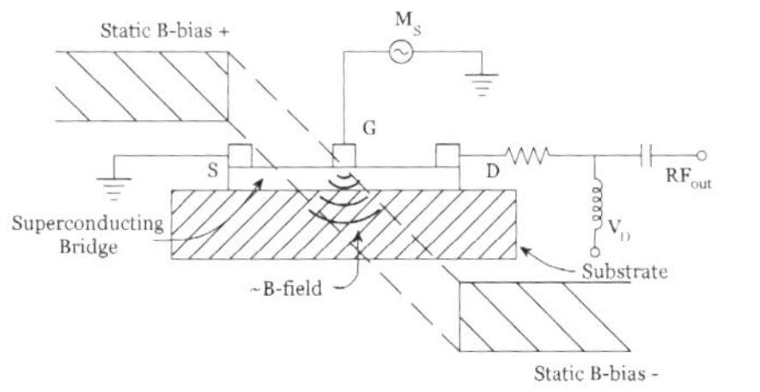

Fig. 1. Operating principle of an MET. A time varying magnetic field applied at the gate electrode (G) modulates the conductivity of the superconducting bridge.

voltage v_{DS} and the magnetic field applied to the bridge H_a.

$$i_D = f\left(H_a, v_{DS}\right) = b_m H_a + \frac{1}{r_d} v_{ds} \quad (1)$$

where

$$b_m = \left.\frac{\partial i_D}{\partial H_a}\right|_{V_{DS}} \approx \left.\frac{\Delta i_D}{\Delta H_a}\right|_{V_{DS}} = \left.\frac{i_d}{H_a}\right|_{V_{DS}} \quad (2)$$

is the magnetic transconductance. The magnetic drain resistance r_d can be defined as

$$r_d = \left.\frac{\partial v_{DS}}{\partial i_D}\right|_{h_a} \approx \left.\frac{\Delta v_{DS}}{\Delta i_D}\right|_{H_a} = \left.\frac{v_{ds}}{i_d}\right|_{H_a} = \frac{1}{\sigma_N}\frac{\delta L}{W\lambda_L^2}\left(\frac{H_a}{H_c}\right)^2. \quad (3)$$

Considering the gate current noise is uncorrelated with the drain current noise, the MET minimum noise temperature is given by[2]

$$T_{\min} = 4\left(\frac{f}{f_T}\right)^2 \frac{r_{gs}}{r_{ds}} T \quad (4)$$

where f is the operating frequency (Hz), f_T is the upper cut-off frequency, $r_{ds} = \frac{1}{\sigma_N}\frac{\delta L}{W\lambda_L^2}\left(\frac{H_a}{H_c}\right)^2$ and $r_{gs} = 2\pi f L_k$.

For MET, the upper cut-off operating frequency is dependent on Cooper pair relaxation time τ_0. In case of Niobium[3],

$$\tau_0 = \frac{\pi\hbar}{8k_B(T - T_{c0})} = 8.33 \;\; 10^{-13} \text{ s} \quad (5)$$

*Submitted 01 June 2020
Siddhartha Sirsi is a Graduate Research Assistant in the Department of Electrical and Computer Engineering, University of Arizona, Tucson, AZ 85721 USA (ssirsi@email.arizona.edu)

Christopher K. Walker is a Professor of Astronomy at the University of Arizona, Tucson, AZ 85718 USA (cwalker@as.arizona.edu).

$$f_t = \frac{1}{\tau_0} = 1.2\times10^{12} Hz \quad 1.2\ \text{THz} \qquad (6)$$

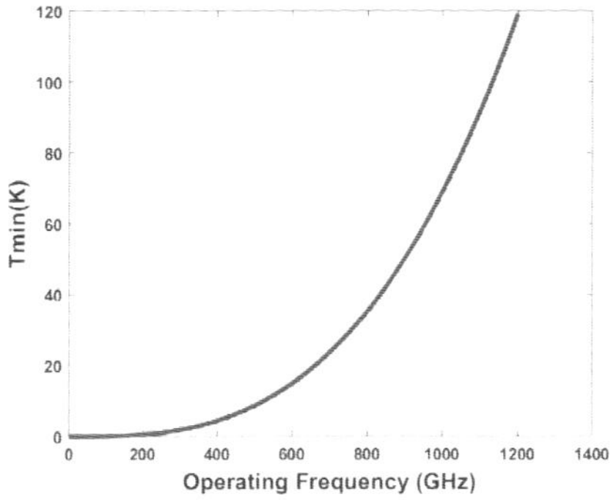

Fig. 2. Expected minimum MET noise temperature vs range of operating frequency. At $f_T = 1.2\ THz$, $T_{\min} = 120$ K.

II. Test Set-Up & Experimental Results

The MET is bolted on to the cold plate inside the helium cryostat. Voltage biasing is provided using a bias tee connected through the pre-amplifier to the SIS & Magnet Bias Box. Backing pump and Turbo pump are used to pull vacuum in the cryostat. The cold head connected to a compressor is then employed to cool the cold plate down to 90 K. Cryostat is then filled with liquid helium. Once the cold plate is cooled down to 4.4 K, the magnet is moved into position and Voltage-Current characteristics of the MET is recorded.

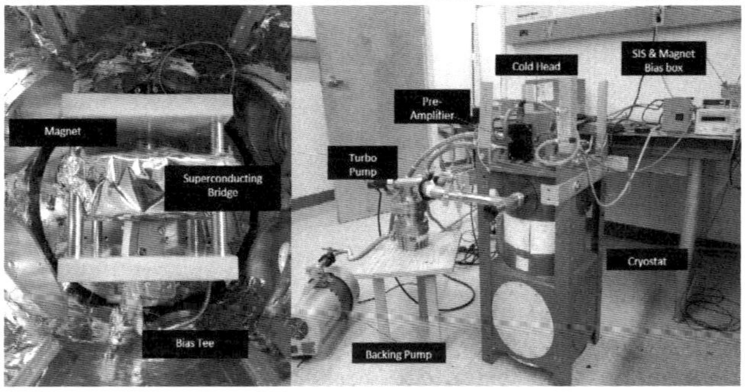

Fig. 3. Left: Prototype MET inside of cryostat. Right: MET Experimental Set-up

A pulse width modulator is used to modulate the current supply to the solenoid at low frequency (around 4-6 Hz). The resulting B_{sig} modulates the conductivity of MET. SIS & Magnet Bias box is used to supply V_{DS} and read the voltage equivalent of I_{DS}. A lock-in amplifier is used to pick up the voltage equivalent of I_{DS}. An Arduino board is programmed to read in this voltage and convert it to I_{DS}.

In order to test the prototype MET it is first biased to its operating point using the neodymium toroidal magnet. Fig. 3 is a plot of the lock-in amplifier output versus time as the position of the toroidal magnet is varied. Since the lock-in amplifier output is proportional to I_{DS}, the plot shows the ability to magnetically bias the MET to its proper operating point. Once at its operating point, the conductivity of the superconducting bridge is then modulated by applying a signal current to the solenoid, I_{sol}.

As expected, when the toroidal magnet is out of position, such that the MET is no longer at its operating point, varying I_{sol} does not have an effect on I_{DS}. Fig. 4 indicates the conductivity of the superconducting bridge is being modulated via B_{sig} by varying I_{sol}. Increasing I_{sol} leads to increase in B_{sig}, which results in a bigger change in I_{DS}.

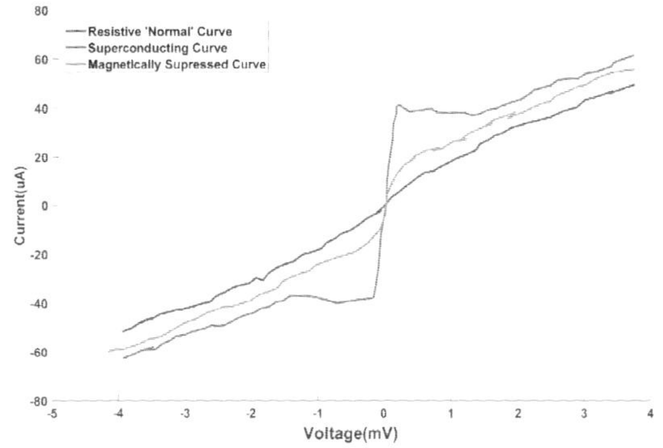

Fig. 3. I-V Curves showing intermediate states of MET as a function of magnetic field - a necessary step in validating MET performance.

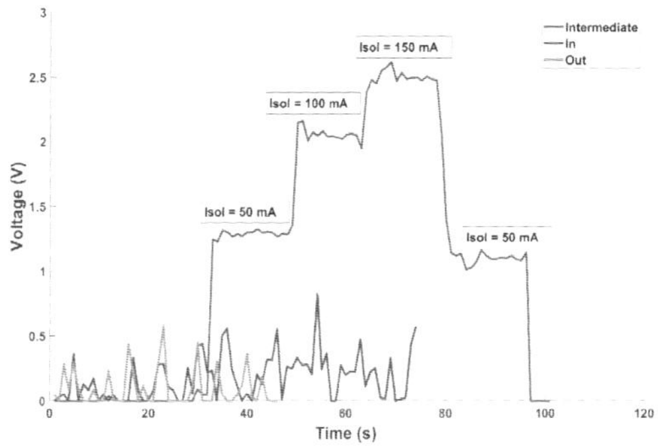

Fig. 4. Lock-In Amplifier output representing the voltage equivalent of I_{DS} for different values of solenoid supply current I_{sol}.

III. Conclusion

In this paper we have introduced the concept of the Meissner Effect Transistor (MET) and presented its theory of operation. Initial test results are shown to be consistent with the underlying theory. Further characterization of MET performance compared to theory will require a dedicated device run and an improved test set-up.

References

[1] Millman, J. and Halkias, C., "Integrated Electronics: Analog and Digital Circuits and Systems", pub. McGraw-Hill (New York), p. 318 (1972)..

[2] K. O. N. Micro, "Noise Parameters," *Electron. Noise Interf. Signals*, no. 9, pp. 69–112, 2006, doi: 10.1007/3-540-26510-4_4.

[3] M. V. Ramallo *et al.*, "Relaxation time of the Cooper pairs near Tcin cuprate superconductors," *Europhys. Lett.*, vol. 48, no. 1, pp. 79–85, 1999, doi: 10.1209/epl/i1999-00117-6.

Design Concept of W-band Multibeam Receiver for the SRT

Alessandro Navarrini, Luca Olmi, Renzo Nesti, Pasqualino Marongiu, Pierluigi Ortu, Luca Cresci, Andrea Orlati, Alessandro Scalambra, and Alessandro Orfei

***Abstract*—We describe the preliminary design concept of a W-band multibeam receiver for the Gregorian focus of the Sardinia Radio Telescope (SRT), a new general purpose fully steerable 64-m diameter antenna located on the Sardinia island, Italy, managed by the Italian National Institute for Astrophysics (INAF).**

The goal specifications of the W-band receiver foresee a 4×4 array of dual-polarization feed-systems utilizing waveguide Orthomode Transducers (OMTs) and low-noise amplifiers (LNAs) cryogenically cooled at ≈20 K, all designed to cover the 70-116 GHz RF band. The instrument concept utilizes a dual sideband separation (2SB) downconversion mixing scheme delivering two 8 GHz-wide Intermediate Frequency (IF) bandwidths (across 4-12 GHz), the USB and the LSB, for each of the two polarization channels of each of the pixels.

The instrument is being procured by INAF through an international call for bid. In support of such procurement, we set the instrument minimum requirements and selected few possible architectures. Then, we conducted an advanced feasibility study of the receiver that allowed us to define a preliminary design of the full receiver for its minimum and goal requirements. In particular, we studied the following receiver subsystems: the optics, the cryogenic modules, the cryostat, the down-conversion and calibration systems, the mechanical derotator (to track the parallactic angle), and all of its essential subsystems. Here, we present the main requirements and the preliminary design of the W-band multibeam receiver, which is set to deliver state-of-the-art performance and will allow to conduct single-dish high-sensitivity large-scale surveys radio astronomy observations with the SRT.

***Index Terms*—Low-Noise Amplifiers, Mixers, Radio astronomy, Receivers, Telescope.**

I. Introduction

THE Sardinia Radio Telescope (SRT, www.srt.inaf.it) is a new general purpose fully steerable 64-m diameter radio telescope designed to operate with high efficiency across the 0.3-116 GHz frequency range [1]. Since December 2018 the telescope, shown in Fig. 1, has been opened to the international community to carry out radio astronomy observing programs using an initial set of receivers covering four RF bands across 0.3-26.5 GHz [2]-[6]. The SRT operates in single-dish (continuum, full Stokes and spectroscopy), Very Long Baseline Interferometry (VLBI) and Space Science modes and it has been successfully used also for space-debris detection and Sun observations.

The telescope optical design is based on a quasi-Gregorian configuration with shaped 64-m diameter primary (M1) and shaped 7.9-m diameter secondary (M2) reflectors to minimize spillover and standing waves (Fig. 2). The primary mirror utilizes an active surface with 1116 electromechanical actuators to compensate the gravitational deformation in real-time. The actuators are also used to convert the shaped surface of the primary mirror to a parabolic profile during primary focus observation.

To extend the current capabilities of the telescope to high frequencies, INAF aims at upgrading the metrology system and at developing and/or procuring a new set of Front-Ends to be installed at its Gregorian focus (Fig. 3).

One of such new instruments is a W-band multibeam heterodyne receiver, whose specifications and design concept are described in the following sections.

Fig. 1. The 64-m diameter Sardinia Radio Telescope (SRT), Sardinia island, Italy.

This paper was submitted for review on June 1st, 2020.

All authors are with the National Institute for Astrophysics (INAF), Italy.

Alessandro Navarrini, Pasqualino Marongiu and Pierluigi Ortu are with INAF-Astronomical Observatory of Cagliari, 09047 Selargius, Italy (e-mails: alessandro.navarrini@inaf.it, pasqualino.marongiu@inaf.it, pierluigi.ortu@inaf.it).

Luca Olmi, Renzo Nesti and Luca Cresci are with INAF-Astrophysical Observatory of Arcetri, 50125 Florence, Italy (e-mails: luca.olmi@inaf.it, renzo.nesti@inaf.it, luca.cresci@inaf.it).

Andrea Orlati, Alessandro Scalambra and Alessandro Orfei are with INAF-Institute of Radio Astronomy, 40129 Bologna, Italy (e-mails: andrea.orlati@inaf.it, alessandro.scalambra@inaf.it, alessandro.orfei@inaf.it).

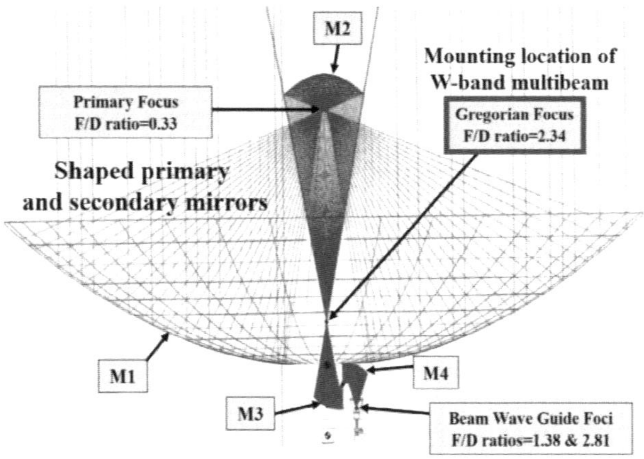

Fig. 2. Optical configuration and ray tracing of the SRT showing the 64-m diameter primary (M1), the 7.9-m diameter secondary (M2), and two additional Beam Waveguide (BWG) mirrors (M3 and M4). The primary and secondary mirrors are shaped. The W-band multibeam receiver will be installed at the Gregorian focus (F/D=2.34).

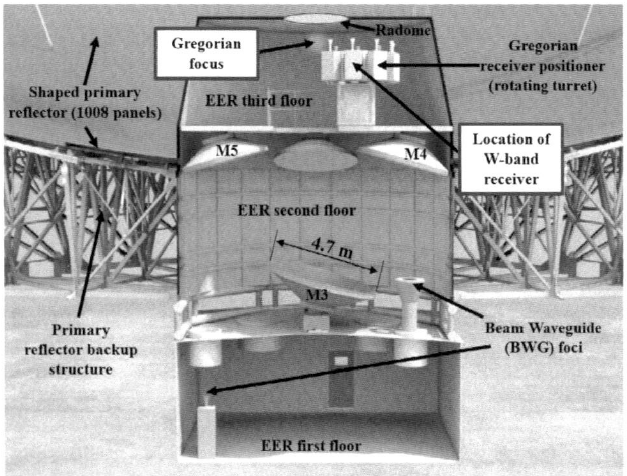

Fig. 3. Cross-cut of the SRT telescope showing details of the three-floor Elevation Equipment Room (EER). The W-band multibeam receiver will be installed on the Gregorian receiver positioner at the Gregorian focus, located on the EER third floor. The image also shows four of the Beam Waveguide (BWG) mirrors on the EER second floor as well as the BWG foci, one of which on the EER first floor.

II. W-BAND MULTIBEAM RECEIVER SPECIFICATIONS

The receiver will be installed at the Gregorian focus of the SRT and must be designed to provide high-efficiency illuminations of the antenna shaped optics for all of its pixels across the 75-116 GHz band, with goal band 70-116 GHz. The minimum requirements and the goal requirements of the W-band receiver are listed in Table I. The instrument must be capable to observe weak radio astronomy sources with high-sensitivity (receiver noise of less than 60 K) but also the strong signals from the Sun without saturating the receiver chains. The array receiver shall be based on a minimum of nine dual linear polarization feeds in a square 3×3 configuration. The goal is an array of 16 feeds in a square 4×4 configuration for the 70-116 GHz band. The receiver must be designed to provide high mapping efficiency by optimizing the geometry and the separation between the projected beams on the sky. The aperture efficiency η_{eff} of the SRT antenna illuminated by all feeds, including the ones with the largest offset from the optical axis, shall be no less than 0.50 (50%) at all frequencies across the full RF band, i.e. $\eta_{eff}=\eta_t\,\eta_s\,\eta_p \geq 0.50$, where η_{eff} is the product of taper efficiency η_t, spillover efficiency η_s and polarization efficiency η_p, (the other contributions to the aperture efficiency, e.g. Ruze, focus, radiation efficiency and blockage, are considered to be unitary, see also [7]-[8] for definitions).

TABLE I
MINIMUM AND GOAL REQUIREMENTS OF THE W-BAND MULTIBEAM RECEIVER

RF band	75-116 GHz (goal 70-116 GHz)
Polarization properties	Two orthogonal linear with OMTs
Number of pixels and array configuration	Square or alternative formats. Minimum 9. Goal 16
Antenna aperture efficiency	$\eta_{eff}=\eta_t\,\eta s\,\eta_p \geq 0.50$ across the RF band for all feeds
Technology	HEMT LNAs
Downconversion scheme and IF band	2x6GHz SB (USB+LSB), IF band 4-10GHz. Goal 2x8GHz SB, IF band 4-12 GHz
Maximum IF outputs	38 for any number of pixels. A 4x4 array must have a maximum of 32xIF outputs.
Mechanical derotator	Yes, to track the parallactic angle
Array calibration	Single ≈293 K calibration load
Solar observations	Yes, with switchable filter/attenuator placed in front of vacuum window
Noise temperature	$T_{SSB} \leq 60$ K over 80% of RF band
Image band suppression	$Ri \geq 10$ dB
Local Oscillator signal	Tunable via high phase stability synthesizers or YIG oscillator
Cryocooler	Two cryogenic stages: S1<80 K; S2<20 K
Overall diameter (mm)	< 800
Height (mm)	≤ 2465
Weight (kg)	< 250

The receiver shall include a calibration system with at least one room temperature (≈293 K) calibration load.

III. RECEIVER ARCHITECTURE

A possible architecture of the receiver is shown in Fig. 4.

The dual linear polarization shall be achieved by means of dual-polarization feed systems where each array element

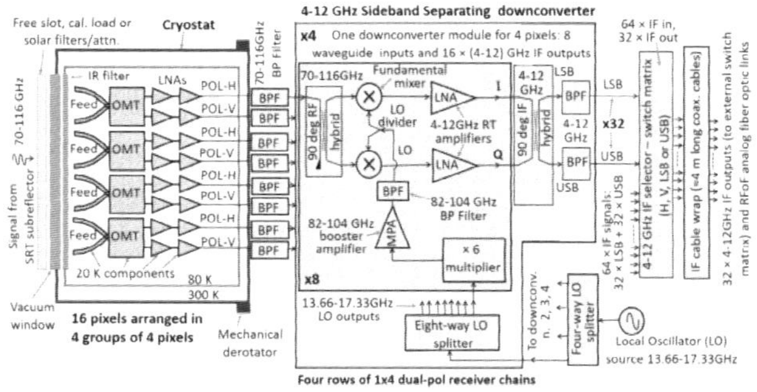

Fig. 4. Possible W-band multibeam receiver architecture in 4×4 configuration covering 70-116 GHz. It shows the calibration load and solar filters/attenuator at its input, the cryostat with the cryogenic components, the room temperature sideband separating down converters, the LO distribution system and the "internal" IF switch matrix to select the IF signals to deliver to the backend. The receiver includes a derotator mechanically attached to the cryostat.

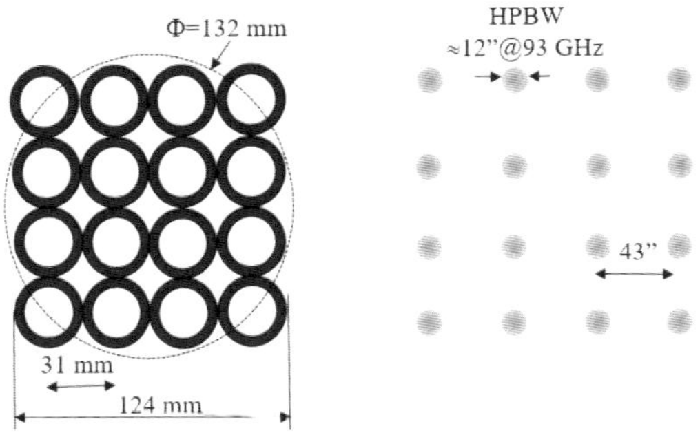

Fig. 5. Array feed configuration and spacing on the Gregorian focal plane (left) and projected beams on the sky (right) for a 4x4 square format solution.

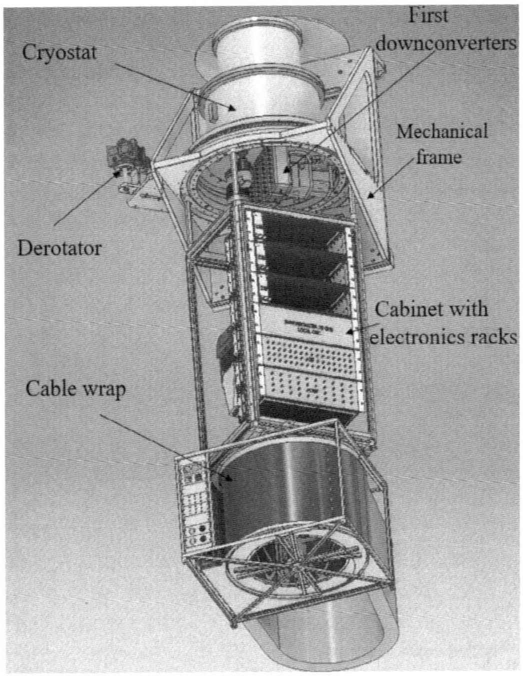

Fig. 6. 3D sketch of the W-band multibeam receiver for the Gregorian focus of the SRT. View of the full instrument showing the cryostat, the downconverters, the mechanical derotator, the cabinet with electronics rack and the mechanical frame for mounting on the Gregorian receiver positioner.

employs a cascade of corrugated feed horn, OMTs and High Electron Mobility Transistor (HEMT) LNAs cryogenically cooled at ≈20 K inside a cryostat. The downconverters for each of the receiving chains, utilizing a tunable Local Oscillator (LO) signal distributed to each of the elements, can be located at cryogenic temperature or at room temperature, inside or outside the cryostat.

The downconverter shall provide a minimum of 12 GHz band of Radio Frequency (RF) sky coverage for each polarization of each of the pixels (goal 16 GHz) and can utilize either a Single Side Band (SSB) or a dual Sideband Separation (2SB) mixing scheme. Fig. 4 shows a possible example of receiver architecture based on 2SB scheme delivering an IF band of 4-12 GHz per sideband (goal specification).

Due to the shaped configuration of the SRT the usable focal plane area at the Gregorian focus is limited to feed horns whose axis are placed within a radius of ≈65 mm from the telescope optical axis. To be confined within a ≈130 mm diameter, a 9 element dual polarization focal plane array in 3×3 configuration can adopt a feed spacing of ≈45 mm, although shorter spacing would be desirable. Instead, a dual polarization array in 4×4 configuration necessarily requires to design cryogenic modules with smaller footprint size where the feed spacing is ≈31 mm. The module would have to adopt OMT waveguide outputs and LNA inputs with non-standard miniaturized waveguide flanges. A possible configuration of the array with 4×4 feeds on the SRT Gregorian focal plane and the corresponding beams projected on the sky are shown in Fig. 5. The example refers to an array of feeds placed on the Gregorian focus without reimaging optics.

We made a preliminary 3D design concept of the full receiver, illustrated in Fig. 6, corresponding to the receiver architecture shown in Fig. 4. The 3D sketch includes the cryostat with commercial cryocooler and first downconverters attached to its backplate, the mechanical frame to mount the instrument on the Gregorian receiver positioner (see Fig. 3), a cabinet with electronics racks for biasing the LNAs and for monitoring and control of the receiver subsystems, a mechanical derotator to maintain the parallactic angle during source tracking with associated cable wrap, a vacuum pump with remotely controlled vacuum valve, the switching mechanism for selecting the calibrator or the solar filter/attenuator in front the vacuum windows and all other minor but necessary accessories.

IV. Conclusions

We presented the specifications and a preliminary design concept for a W-band multibeam receiver to be installed at the Gregorian focus of the SRT. The instrument is being procured by INAF through a call for bid.

References

[1] I. Prandoni et al. "The Sardinia Radio Telescope: From a Technological Project to a Radio Observatory," *Astronomy & Astrophysics*, 2017, A&A, 608, A40.

[2] A. Navarrini et al. "The Sardinia Radio Telescope Front-Ends," *Proceedings of 27th Int. Symp. On Space THz Tech. ISSTT2016*, Nanjing, China, 13-15 Apr. 2016.

[3] G. Valente, I. Pisanu, P. Bolli, S. Mariotti, P. Marongiu, A. Navarrini, R. Nesti, A. Orfei, *The dual-band LP feed system for the Sardinia Radio Telescope prime focus*, in Proc. SPIE Millimeter, Submillimeter, and Far-Infrared Detectors and Instrumentation for Astronomy V, Vol. 7741, 774126, 2010.

[4] A. Orfei, A. Cattani, M. Poloni, J. Roda, S. Mariotti, A. Maccaferri, A. Orlati, A. Scalambra, *Caratterizzazione del rivcevitore nella banda 5.7-7.7 GHz*, SRT Memo Series GAI04-FR-5.0 May 5th, 2011.

[5] A. Orfei, L. Carbonaro, A. Cattani, A. Cremonini, L. Cresci, F. Fiocchi, A. Maccaferri, G. Maccaferri, S. Mariotti, J. Monari, M. Morsiani, V. Natale, R. Nesti, D. Panella, M. Poloni, J. Roda, A. Scalambra, G. Tofani, *A Multi-Feed Receiver in the 18 to 26.5 GHz Band for Radio Astronomy*, IEEE Antennas Propag. Mag. 52(4), 62, 2010.

[6] A. Navarrini et al. "Front-Ends and Phased Array Feeds for the Sardinia Radio Telescope," *Proceedings of 32nd URSI GASS*, Montreal, 19-26 Aug. 2017.

[7] J. Baars, "Characteristics of a Reflector Antenna: Parameters, graphs and formulae for Cassegrain systems with Mathematica expressions for numerical computation", *ALMA Memo 456*, April 2003.

[8] L. Olmi, P. Bolli, "Ray-tracing and physical-optics analysis of the aperture efficiency in a radio telescope," *Applied Optics*, 46, 4092-4101, 2007.

Developing High-Sensitivity Graphene Terahertz Detectors Through A High-Yield Nanofabrication Process

Panagiotis C. Theofanopoulos[1] and Georgios C. Trichopoulos[1]

Recently it has been demonstrated that graphene can be used to develop large-scale, high-sensitivity coherent detection receivers for astronomy applications in the terahertz bands (300 GHz-5 THz) [1]. Such detectors can be used in large focal plane arrays for high resolution astronomical imaging [2], while offering increased operation bandwidth compared to other superconductors (e.g. niobium nitride– NbN) with lower local-oscillator power requirements, due to the exhibited quantum effects in graphene [1]. However, developing such large-scale graphene sensors is hindered, since graphene fabrication suffers from low yields over large areas [3]. Specifically, graphene adhesion to the carrier substrate (e.g. SiO_2 or Si) is typically weak thus leading to easy delamination of graphene during the nanofabrication process (e.g. development, liftoff, etc.) due to the used chemicals [3].

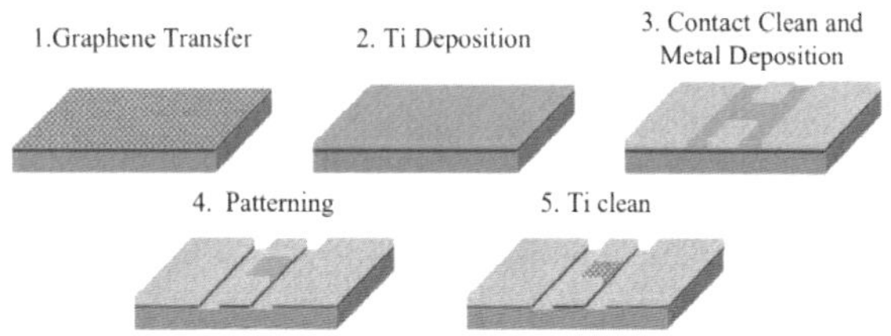

Fig. 1. The proposed nano-fabrication procedure used to develop large-area graphene devices: 1. Transfer graphene on the high-resistivity Si substrate, 2. Deposit a 30 nm Ti sacrificial layer using electron-beam evaporation, 3. Form the metal layer of gold/chrome using photolithography, electron-beam evaporation, and lift-off (before the metal deposition we clean the Ti from the open areas to ensure good metal-graphene contact) 4. Pattern graphene devices using dry etching, 5. Clean the Ti sacrificial layer using wet chemistry.

The scope of our work is to increase the yield of graphene nanofabrication over large areas enabling the development of multi-element graphene sensor arrays for astronomical and other applications. To achieve that we use a sacrificial layer as shown in Fig. 1, that protects the graphene throughout the nanofabrication steps and does not allow the delamination during the wet processes. Exploiting the proposed method, we acquire 92 % yield over a 2 cm × 2 cm aperture, improving the almost zero yield exhibited without the use of the sacrificial layer. The sacrificial layer consists of a 30 nm thick titanium–Ti film and is deposited using electron beam evaporation. The advantages of the proposed method are twofold: 1) we prevent graphene delimitation during the wet processes and 2) Ti helps to retrieve the intrinsic graphene properties by removing any contaminants that adhere to the graphene lattice.

Our fabrication process is verified through Raman spectroscopy and graphene sheet resistance is measured in the 220-330 GHz bands. The test fixture is comprised of a graphene loaded coplanar waveguide (CPW) transmission lines and with the use of RF contact probes we measure the scattering parameters of the devices. Afterwards, the graphene sheet impedance is extracted, and the results are verified with already published showing good agreement. Moreover, the graphene sheet impedance is measured for various biasing voltages and the acquired mobility changes are in accordance with the existing literature, revealing that the proposed nanofabrication process has not harmed the 2D graphene layer.

During the conference, the proposed graphene high-yield fabrication method will be presented alongside the contact probe measurements process.

References

[1] Lara-Avila, S., Danilov, A., Golubev, D. et al. "Towards quantum-limited coherent detection of terahertz waves in charge-neutral graphene," Nat Astron 3, 983–988 (2019).

[2] P. F. Goldsmith, "Sub-millimeter heterodyne focal-plane arrays for high-resolution astronomical spectroscopy," in URSI Radio Science Bulletin, vol. 2017, no. 362, pp. 53-73, Sept. 2017.

[3] N. C. Wang, E. A. Carrion, M. C. Tung, and E. Pop, "Reducing graphene device variability with yttrium sacrificial layers," Appl. Phys. Lett., vol. 110, no. 22, p. 223106, 2017.

[1] School of Electrical, Computer, and Energy Engineering, Arizona State University, Tempe, AZ, USA

A small satellite with a dual-frequency heterodyne spectrometer for the detection of atomic oxygen in the atmosphere of Earth

H. Richter[1], J. Hildebrandt[2], T. Roth[2], M. Lengowski[2], C. Philpot[3], A. Braukhane[3], T. Delovski[3], M. Wienold[1], S. Klinkner[2], H.-W. Hübers[1,4]

Atomic oxygen (OI) is the main component of the mesosphere and lower thermosphere (MLT, altitude approx. 80-300 km) of the Earth. It is generated through photolysis of molecular oxygen or ozone by ultraviolet radiation from the Sun. The photochemistry and the energy balance of the MLT are governed by OI. In addition, it is a tracer for dynamical motions in the MLT. OI is extremely difficult to measure with remote sensing techniques, since it has not many optically active transitions. We have measured the fine-structure transition of OI at 4.7448 THz using the GREAT heterodyne spectrometer on board of SOFIA, the Stratospheric Observatory for Infrared Astronomy. This is the first measurement of the OI emission line shape at 4.7448 THz. This method enables the direct measurement of OI without involving photochemical models and agrees within 10% with recent atmospheric models. In contrast, OI concentrations derived from satellite instruments are not derived from direct observations of OI and involve complex photochemical models. This is where a small satellite mission could be beneficial. The derived concentrations are up to 100% larger than predicted by atmospheric models. Therefore, we propose to measure the fine structure transitions of OI at 2.06 and 4.7 THz with a dual-frequency heterodyne spectrometer on a small satellite. This satellite mission, called OSAS (Oxygen Spectrometer for Atmospheric Science) will yield more accurate results than previous missions, global coverage and annual variations of OI.

As a first step towards realization of OSAS, a small satellite study has been performed based on Concurrent Engineering methods. The proposed science payload is a dual frequency heterodyne spectrometer (Fig 1). The low-frequency channel is centered at the 2.06 THz fine-structure transition of OI and the high-frequency channel is centered at its 4.7-THz fine-structure transition. The low-frequency channel has a Schottky diode mixer with a multiplied microwave oscillator as local oscillator (LO). The high-frequency channel is based on a Schottky diode mixer and a quantum-cascade laser (QCL) as LO. The backend is a digital fast Fourier spectrometer. To provide complete OI mapping of the Earth plus characterization of particular areas of interest, the spacecraft design options allow the observation of the MLT at different local times during the time of the mission. Therefore, a high inclination low earth orbit, preferably a sun synchronous orbit (SSO), is intended as initial orbit. In case of a SSO a propulsion system adapts the orbit to observe the MLT at different local times. A total mission duration of 24 months is planned. Five solar array panels will provide up to 260 W electrical power required by the entire spacecraft for this challenging mission. The total mass of the spacecraft (Fig. 2) has been estimated at approx. 240 kg with the 50 kg OSAS payload.

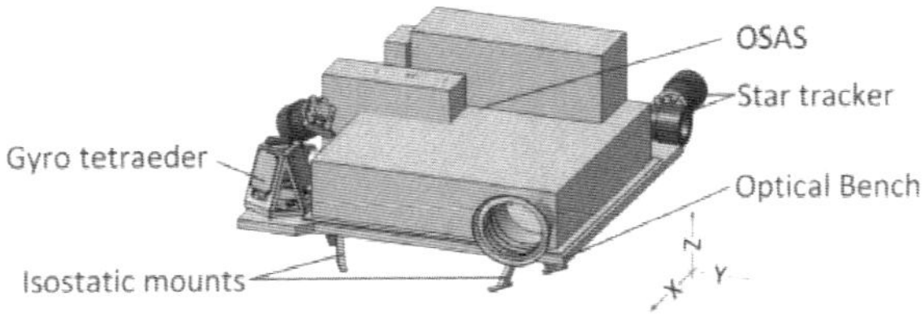

Fig. 1: Scheme of the optical bench with OSAS payload.

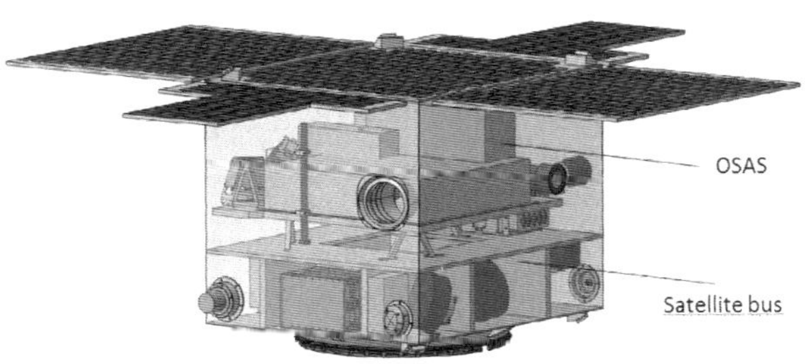

Fig. 2: Scheme of the satellite (deployed).

[1] German Aerospace Center (DLR), Institute of Optical Sensor Systems, Rutherfordstr. 2, 12489 Berlin, Germany
[2] University of Stuttgart, Institute of Space Systems, Pfaffenwaldring 29, 70569 Stuttgart, Germany
[3] German Aerospace Center (DLR), Institute of Space Systems, Robert-Hooke-Str. 7, 28359 Bremen, Germany
[4] Humboldt-Universität zu Berlin, Department of Physics, Newtonstr. 15, 12489 Berlin, Germany

SuperSpec: On-Chip Direct-Detection Spectroscopy, Preparing for LMT Campaign

Joe Redford[1], Steve Hailey-Dunsheath[1], H.G. (Rick) LeDuc[2], Reinier Janssen[2], C.M. (Matt) Bradford[2,1], Ryan McGeehan[3], Kirit Karkare[3], Erik Shirokoff[3], Pete Barry[4], Jordan Wheeler[5], Jason Glenn[6], Phil Mauskopf[7], Carole Tucker[8], Ryley Hill[9], and Scott Chapman[9]

Wideband, direct-detection spectroscopy in the far-IR through millimeter is a compelling scientific opportunity on platforms ranging from ground-based to cryogenic orbital. Large-format spectroscopy in these bands is challenging, however, because it requires large arrays of sensitive detectors, and because conventional optical spectrometer designs are large and do not scale up gracefully.

Fig. 1. *Top*: 300-channel SuperSpec chip covering the 200-300 GHz band. The feedline runs from right to left (~1/3 down from the top). The vertical structures are the KID capacitors. *Bottom*: Cold box of the demonstration instrument. It houses 6 chips to couple both polarizations of 3 beams (grid polarizer not shown here); they will be steered to the source with a chopping mirror at a pupil in the warm relay optics.

To address these challenges, we have developed a superconducting on-chip filterbank, SuperSpec. Radiation propagates on niobium on-silicon-nitride microstrip, and encounters a series of resonant filters, each coupling a narrow band to titanium-nitride kinetic inductance detector (KID). As Figure 1 shows, a full wideband spectrometer fits on a silicon chip a few square cm in size.

In previous 50-channel prototypes covering the 240 to 265 GHz band, we demonstrated good filterbank performance. and a limiting detector noise equivalent power of 7×10^{-19} W $Hz^{-1/2}$, comfortably photon-background-limited for any ground- or suborbital application. We have developed and are now producing full band (~200-300 GHz) 300-channel (Fig. 1) and 110-channel chips for scientific use.

We are preparing to deploy 6 chips in a demonstration instrument (Fig. 1, bottom) at the Grand Telescopio Millimetrico (GTM/LMT) on the 15,092' Sierra Negra in Mexico. The spectrometer chips are cooled to 220 mK, and each uses a single readout circuit operating between 100-450 MHz based on with ROACH-2 electronics. The instrument will be installed this summer with scientific observations in the fall. We will target dusty galaxies at intermediate and high redshifts in CO and [CII], respectively.

Future prospects for SuperSpec include steered multi-object spectrometers and large 2-D imaging spectrometers.

References

[1] J. Redford et al., The design and characterization of a 300 channel, optimized full-band millimeter filterbank for science with SuperSpec, Proceedings of the SPIE 10708, 107081O (2018).

[2] K. Kirkare et al., Full-array noise performance of deployment-grade SuperSpec mm-wave on-chip spectrometers, Journal of Low Temperature Physics, in press (2020).

[1] California Institute of Technology, Pasadena, CA 91125 USA.
[2] Jet Propulsion Laboratory, Caltech, Pasadena, CA 91109 USA.
[3] University of Chicago, Chicago, IL 60637 USA.
[4] Argonne National Laboratory, Lemont, IL 60439 USA.
[5] National Institute of Standards and Technology, Boulder, CO 80305 USA.
[6] NASA Goddard Space Flight Center, Greenbelt, MD 20771 USA.
[7] Arizona State University, Tempe, AZ 85281 USA.
[8] Cardiff University, Cardiff CF10 3AT, United Kingdom.
[8] Cardiff University, Cardiff CF10 3AT, United Kingdom
[9] University of British Columbia, Vancouver, BC V6T 1Z4, Canada.

NOTES:

List of registered participants

First Name	Last Name	Email Address	Institution
Lingzhen	Zeng	lingzhen@cfa.harvard.edu	Center for Astrophysics | Harvard & Smithsonian
Edward	Tong	etong@cfa.harvard.edu	Center for Astrophysics | Harvard & Smithsonian
Wenlei	Shan	wenlei.shan@nao.ac.jp	National Astronomical Observatory of Japan
John	Garrett	john.garrett@cfa.harvard.edu	Center for Astrophysics | Harvard & Smithsonian
Boris	Karasik	boris.s.karasik@jpl.nasa.gov	JPL
Hiroshi	Matsuo	h.matsuo@nao.ac.jp	National Astronomical Observatory of Japan
Richard	Wylde	r.wylde@terahertz.co.uk	Thomas Keating Ltd.
Hajime	Ezawa	h.ezawa@nao.ac.jp	National Astronomical Observatory of Japan
Darren	Dowell	charles.d.dowell@jpl.nasa.gov	JPL
Yoshihisa	Irimajiri	irimaji@nict.go.jp	NICT
Pavel	Yagoubov	pyagoubo@eso.org	European Southern Observatory
Jan	Barkhof	j.barkhof@sron.nl	NOVA
Andrey	Baryshev	andrey@astro.rug.nl	Kapteyn Astronomical Institute/NOVA
Raymond	Blundell	rblundell@cfa.harvard.edu	Center for Astrophysics | Harvard & Smithsonian
Sergey	Cherednichenko	serguei@chalmers.se	Chalmers University of Technology
Vladimir	Drakinskiy	vladimir.drakinskiy@chalmers.se	Chalmers University of Technology
Ronald	Hesper	r.hesper@astro.rug.nl	Kapteyn Astronomical Institute, University of Groningen
N.	Honingh	honingh@ph1.uni-koeln.de	University Cologne
Takafumi	Kojima	t.kojima@nao.ac.jp	National Astronomical Observatory of Japan
Patrick	Pütz	patrick.puetz@radiometer-physics.de	Radiometer Physics GmbH
Sho	Masui	s_s.masui@p.s.osakafu-u.ac.jp	Osaka Prefecture University
Paul	Grimes	pgrimes@cfa.harvard.edu	Smithsonian Astrophysical Observatory
Christophe	Risacher	risacher@iram.fr	IRAM
Christine	Chen	Christine.p.chen@jpl.nasa.gov	JPL
Heinz-Wilhelm	Hübers	heinz-wilhelm.huebers@dlr.de	German Aerospace Center (DLR), Institute of Optical Sensor Systems
Theodore	Reck	theodore.reck@vadiodes.com	Virginia Diodes, Inc.
Changyun	Yoo	cyoo@ucsb.edu	University of California Santa Barbara
Jeffrey	Hesler	hesler@vadiodes.com	Virginia Diodes, Inc.
Eduard	Driessen	driessen@iram.fr	IRAM
Kitti	Ratter	kitti.ratter@physics.ox.ac.uk	University of Oxford
Alessandro	Traini	alessandro.traini@physics.ox.ac.uk	University of Oxford
Boon Kok	Tan	boonkok.tan@physics.ox.ac.uk	University of Oxford
Tom	Crowe	crowe@vadiodes.com	Virginia Diodes, Inc.
Bert	Hawkins	bhawkins@nrao.edu	NRAO
Alessandro	Navarrini	alessandro.navarrini@inaf.it	INAF (National Institute for Astrophysics)
Reinier	Janssen	rjanssen@jpl.nasa.gov	NASA JPL
Ming-Jye	Wang	mingjye@asiaa.sinica.edu.tw	Institute of Astronomy and Astrophysics, Academia Sinica

Deacon	Nemchick	deacon.j.nemchick@jpl.nasa.gov	JPL, California Institute of Technology
Adrian	Gomez Torrent	adriango@kth.se	KTH Royal Institute of Technology
Rong	Nie	rongnie2@illinois.edu	University of Illinois
Faouzi	Boussaha	faouzi.boussaha@obspm.fr	CNRS-GEPI
Keara	Carter	keara.carter@cfa.harvard.edu	Center for Astrophysics | Harvard & Smithsonian
Paul	Goldsmith	paul.f.goldsmith@jpl.nasa.gov	JPL
Imran	Mehdi	imran.mehdi@jpl.nasa.gov	JPL
Kamaljeet	Saini	ksaini@nrao.edu	NRAO
Anthony	Cortez	acort038@ucr.edu	University of California Riverside
Yijing	Deng	ydeng2@nd.edu	University of Notre Dame
Yosuke	Murayama	s1830079@s.tsukuba.ac.jp	University of Tsukuba / NAOJ
David	Burghoff	dburghoff@nd.edu	University of Notre Dame
Vincent	Desmaris	vincent.desmaris@chalmers.se	GARD/Chalmers University
Jian-Rong	Gao	j.r.gao@tudelft.nl	SRON & TU Delft
Subash	Khanal	subash.khanal@jpl.nasa.gov	NASA JPL
Adrian	Tang	ajtang@jpl.nasa.gov	JPL/UCLA
Mohamed Aniss	Mebarki	aniss@chalmers.se	Chalmers
Patricio	Mena	fpmena@uchile.cl	Universidad de Chile
Omkar	Pradhan	omkarpradhan@gmail.com	NASA JPL
Christopher	Curwen	chris.a.curwen@jpl.nasa.gov	JPL
Martin	Wienold	martin.wienold@dlr.de	German Aerospace Center (DLR)
Sofia	Rahiminejad	sofia.rahiminejad@jpl.nasa.gov	JPL
William	Deal	billdeal@live.com	Northrop Grumman
Eugene	Lauria	glauria@email.arizona.edu	Arizona Radio Observatory
George	Reiland	gpr1@email.arizona.edu	Arizona Radio Observatory
David	Allee	allee@asu.edu	School of Electrical Comp. & Energy Eng., ASU
Fei	Yang	yangfei@seu.edu.cn	Southeast University
Zach	Griffith	zach.griffith@teledyne.com	Teledyne Scientific
Joseph	Lambert	jlambert@nrao.edu	NRAO
Chris	Groppi	cgroppi@asu.edu	Arizona State University
Phil	Mauskopf	Philip.Mauskopf@asu.edu	Arizona State University
Sean	Bryan	Sean.A.Bryan@asu.edu	Arizona State University
Saeed	Zeinolabedinzadeh	Saeed.Zeinolabedinzadeh@asu.edu	Arizona State University
George	Trichopolous	gtrichop@asu.edu	Arizona State University
Lucas	Ziomek	lucas.ziomek@asu.edu	Arizona State University
Jonathan	Hoh	Jonathan.Hoh@asu.edu	Arizona State University
Cassie	Whitton	Cassandra.Whitton@gmail.com	Arizona State University
Marko	Neric	Marko.Neric@asu.edu	Arizona State University
Adrian	Sinclair	Adrian.Sinclair@asu.edu	Arizona State University
Matthew	Underhill	maunderh@asu.edu	Arizona State University
Hamdi	Mani	hamdi.mani@asu.edu	Arizona State University

Justin	Matthewson	Justin.Mathewson@asu.edu	Arizona State University
Kyle	Massingill	kyle.massingill@asu.edu	Arizona State University
Panos	Theofanopoulos	panos.theofano@gmail.com	Arizona State University
Thomas	Mozdzen	Thomas.Mozdzen@asu.edu	Arizona State University
Farzad	Faramarzi	ffarama1@asu.edu	Arizona State University
Sasha	Sypkens	ssypkens@asu.edu	Arizona State University
Paul	Horton	pahorton@asu.edu	Arizona State University